NEW APPROACHES TO THE IDENTIFICATION OF MICROORGANISMS

TECHNIQUES IN
PURE AND APPLIED MICROBIOLOGY

CARL-GÖRAN HEDÉN, Editor

Karolinska Institute, Stockholm

G. Sermonti, *Genetics of Antiobiotic-Producing Microorganisms*

V. B. D. Skerman, Editor, *Abstracts of Microbiological Methods*

Noel R. Rose and Pierluigi E. Bigazzi, Editors, *Methods in Immunodiagnosis*

Carl-Göran Hedén and Tibor Illéni, Editors, *Automation in Microbiology and Immunology*

Carl-Göran Hedén and Tibor Illéni, Editors, *New Approaches to the Identification of Microorganisms*

Brij M. Mitruka, *Gas Chromatographic Applications in Microbiology and Medicine*

NEW APPROACHES TO THE IDENTIFICATION OF MICROORGANISMS

EDITED BY

CARL-GÖRAN HEDÉN

AND

TIBOR ILLÉNI

Karolinska Institute, Stockholm

A WILEY BIOMEDICAL-HEALTH PUBLICATION

JOHN WILEY & SONS
New York · London · Sydney · Toronto

Copyright © 1975, by John Wiley & Sons, Inc.

All rights reserved. Published simultaneously in Canada.

No part of this book may be reporduced by any means, nor transmitted, nor translated into a machine language without the writen permission of the publisher.

Library of Congress Cataloging in Publication Data:

Main entry under title:
New approaches to the identification of microorganisms.

   (Techniques in pure and applied microbiology)   (A Wiley biomedical-health publication)
   Papers presented at a symposium on rapid methods and automation in microbiology held in Stockholm, June 3–9, 1973, and sponsored by UNESCO, WHO, and the International Organization for Biotechnology and Bioengineering.
      1.   Micro-organisms—Identification—Congresses.
2.   Microbiology—Automation—Congresses.   3.   Microbiology—Technique—Congresses. I. Hedén, Carl Göran, ed. II. Illéni, Tibor, 1937–   ed.   III.   United Nations Educational, Scientific, and Cultural Organization.
      IV.   World Health Organization.   V.   International Organization for Biotechnology and Bioengineering.
VI.   Series.   [DNLM: 1.   Bacteriological techniques—Congresses.   2.   Microbiology—Congresses.   3.   Bacteriological techniques—Laboratory manuals.   4.   Microbiology —Laboratory manuals.   QW25 N532 1973]

QR65.N49   576   74-11484
ISBN 0-471-36746-X

Printed in the United States of America

10 9 8 7 6 5 4 3 2 1

AUTHORS

Hans M. Aus
Institute for Viruses
University of Würzburg
Würzburg, West Germany

Peter H. Bartels
Department of Microbiology
and Optical Sciences Center
University of Arizona
Tucson, Arizona

Stellan Bengtsson
Institute of Medical Microbiology
University of Uppsala
Uppsala, Sweden

Fred-Olof Bergqvist
University Data Center
University of Uppsala
Uppsala, Sweden

G. C. Blanchard
Instrumentation Laboratory, Inc.
Lexington, Massachusetts

E. A. Boling
Boston Veterans Administration
Hospital
Boston, Massachusetts

V. Bonifas
Institute of Microbiology
University of Lausanne
Lausanne, Switzerland

D. F. J. Brown
Clinical Research Centre
Harrow, Middlesex, England

H. Brunner
Boehringer-Mannheim G.m.b.H.
Biochemica-Werk Tutzing
Tutzing, West Germany

H. Bürger
Hygiene Institute and
Medical-Research
Departments of the
University of Marburg
Marburg, West Germany

Paxton Cady
Bactomatic, Inc.
Palo Alto, California

Ann Dahlbäck
Department of Bacteriology
National Bacteriological Laboratory
Stockholm, Sweden

S. R. Das
Division of Microbiology
Central Drug Research Institute
Lucknow, India

J. De Ley
Laboratory of Microbiology and
Microbiological Genetics
Faculty of Sciences
State University
Ghent, Belgium

G. Demierre
Institute of Microbiology
University of Lausanne
Lausanne, Switzerland

Akinyele Fabiyi
Virus Research Laboratory
University College Hospital
University of Ibadan
Ibadan, Nigeria

Daniel Y. C. Fung
Department of Microbiology
Pennsylvania State University
University Park, Pennsylvania

Rudolf Gallien
Diagnostic Division
Hoffmann-La Roche AG
Grenzach, West Germany

Donald A. Glaser
University of California
Berkeley, California

H. G. Gyllenberg
Department of Microbiology
University of Helsinki
Helsinki, Finland

Paul A. Hartman
Department of Bacteriology
Iowa State University
Ames, Iowa

Carl-Göran Hedén
Department of Bacteriology
Karolinska Institute and the
Medical Research Council
Stockholm, Sweden

G. Holz
Boehringer -Mannheim G.m.b.H.
Biochemica-Werk Tutzing
Tutzing, West Germany

K. Kersters
Laboratory of Microbiology
and Microbiological Genetics
Faculty of Sciences
State University
Ghent, Belgium

Piet G. Kistemaker
FOM (Institute for Atomic and
Molecular Physics)
Amsterdam, The Netherlands

Donald P. Kronish
Department of Diagnostics Research
Warner-Lambert Research Institute
Morris Plains, New Jersey

J. Leonardopoulus
Department of Microbiology
Faculty of Medicine
National University of Athens
Athens, Greece

Volker ter Meulen
Institute for Viruses
University of Würzburg
Würzburg, West Germany

Henk L. C. Meuzelaar
FOM (Institute for Atomic and
Molecular Physics)
Amsterdam, The Netherlands

Brij M. Mitruka
Department of Research Medicine
University of Pennsylvania
Philadelphia, Pennsylvania Service

Richard Moore
Public Health Laboratory Service
Colindale, London, England

R. S. Moussa
Control Laboratories of
Nestlé Products
Technical Assistance Company
La Tour de Peilz, Switzerland

Thomas R. Neblett
Department of Pathology
Henry Ford Hospital
Detroit, Michigan

S. W. B. Newsom
Sims Woodhead
Memorial Laboratory
Papworth Hospital
Cambridge, England

T. K. Niemelä
Department of Microbiology
University of Helsinki
Helsinki, Finland

Carl-Erik Nord
Department of Bacteriology
National Bacteriological Laboratory
Stockholm, Sweden

Rüknettin Ögütman
Atatürk University Medical School
Department of Microbiology and
Infectious Diseases
Erzurum, Turkey

J. Papavassiliou
Department of Microbiology
Faculty of Medicine
National University of Athens
Athens, Greece

Maarten A. Posthumus
FOM (Institute for Atomic and
Molecular Physics)
Amsterdam, The Netherlands

R. Quast
Hygiene Institute of the University
of Marburg
Marburg, West Germany

O. Ribeiro
Institute of Microbiology
University of Lausanne
Lausanne, Switzerland

W. J. Russell
Instrumentation Laboratory, Inc.
Lexington, Massachusetts

Werner Schneider
University Data Center
University of Uppsala,
Uppsala, Sweden

H.-R. Schulten
Institute for Physical Chemistry
University of Bonn
Bonn, West Germany

Anneke Tom
FOM (Institute for Atomic and
Molecular Physics)
Amsterdam, The Netherlands

Amiram Ur
Division of Bioengineering and
Hospital Infections
Clinical Research Centre
Harrow, Middlesex, England

Torkel Wadström
Department of Bacteriology
National Bacteriological Laboratory
Stockholm, Sweden

George L. Wied
Department of Obstetrics and
Gynecology and Pathology
University of Chicago
Chicago, Illinois

J. F. Zettler
Instrumentation Laboratory, Inc.
Lexington, Massachusetts, and
Boston Veterans Administration
Hospital
Boston, Massachusetts

CONFERENCE PARTICIPANTS

Mohamed Abdou
Boehringer Sohn
Ingelheim, West Germany

Jacques Acar
Hopital St. Joseph
Paris, France

Daniel Amsterdam
Kingsbrook Jewish Medical
Center
Brooklyn, New York

Pertti Arstila
Dept. of Virology
University of Turku
Turku, Finland

Hans Aus
Institut für Virologie
Universität Würzburg
Würzburg, West Germany

Nils Axelsen
Protein Laboratory
Universität of Copenhagen
Copenhagen, Denmark

Åse Bakketun
Norwegian Institute for Water
Research
Oslo, Norway

Zoltán G. Bánhidi
Department of Microbiology
Royal Inst. of Technology
Stockholm, Sweden

Graham Barclay
Hatfield Polytechnic
Hatfield, England

Lothar Behrendt
Farbwerke Hoechst
Frankfurt/M,
West Germany

John E. Benbough
Microbiology Research
Establishment
Salisbury, England

Richard Benn
Department of Bact.
Royal Prince Alfred Hospital
Camperdown, Australia

Theo M. Berg
Organon International B. V.
Oss, The Netherlands

Tom Bergan
Microbiological Institute
University of Tromsö
Tromsö, Norway

Robert Bergquist
Immunological Department
National Bacteriological
Laboratory
Stockholm, Sweden

Bengt Bergrahm
Orion Pharmaceutical
Helsinki, Finland

Ernst Berkhan
Phywe AG
Göttingen, West Germany

Sunil Kumar Biswas
Department of Pathology and
Bacteriology
B.S. Medical College
Bankura, West Bengal, India

Elmar Blatt
SCS Scientific Control Systems
Hamburg, West Germany

Henry Bloom
BioQuest
Cokeysville, Maryland

Philip Blume
Department of Laboratory
Medicine
University of Minnesota
Minneapolis, Minnesota

Arne Bolinder
Technicon AB
Solna, Sweden

Miroslav Bomar
Bundesforschungsanstalt für
Lebensmittelfrischhaltung
Karlsruhe, West Germany

Vallentin Bonifas
Institute of Microbiology
University of Lausanna
Lausanne, Switzerland

Umberto Bonomi
Laboratoria Microbiologia
Istituti Opsitalieri
Verona, Italy

Jerzy Borowski
Department of Microbiology
University Medical School
Bialystok, Poland

Åke Bovallius
The Research Institute of the
National Defenseg
Sundbyberg, Sweden

Niels Brems
A S N Foss Electric
Hilleröd, Denmark

Alan E. Brookfield
Mast Laboratories Ltd.
Liverpool, England

J. E. Brorson
Medicinsk mikrobiologi
Göteborgs Sjukvårdsstyrelse
Göteborg, Sweden

Derek Brown
Clinical Research Centre
Division of Hospital Infection
Harrow, England

Herwig Brunner
Boehringer-Mannheim GmbH
Biochemica-Werk Tutzing
Tutzing, West Germany

Klaus Bryn
Methodology Department
National Institute
for Public Health
Oslo, Norway

Rolf Brönnestam
Clinical Bacteriological
Laboratory
Umeå Hospital
Umeå, Sweden

A. C. J. Burgers
UNESCO
Paris, France

Valerie Nusbaum Bush
Delaware State College
Dover, Delaware

Heinz Burger
Hygiene-Institut der
Universität
Marburg, West Germany

Paxton Cady
Bactomatic, Inc.
Palo Alto, California

Jephta Campbell
Division of Microbiology
U. S. Food and Drug
Administration
Cincinnati, Ohio

Alain Canavate
Lab. Lepetit
Suresnes, France

Torbjörn Caspersson
Karolinska Institute
Stockholm, Sweden

E. Chapelle
Systems Analysis Branch
NASA/GSIC
Greenbelt, Maryland

Olaf Corydon-Petersen
A. S. N. Foss Electric
Hilleröd, Denmark

Dimitro Costin
E. Merck Svenska AB
Stockholm, Sweden

Scott L. Cram
Biophysics and Instrum. Gr.
Los Almos Scientific
Laboratory
Los Alamos, New Mexico

Alice de Czckala
Hoffmann-La Roche, Inc.
Nutley, New Jersey

Karin Dalenius
AB Pripps Bryggerier
Bromma, Sweden

Dan Danielsson
Clinical Bacteriology
Laboratory
The Regional Hospital
Örebro, Sweden

S. R. Das
Central Drug Research
Institute
Lucknow, India

Edgar DaSilva
Department of Microbiology
St. Xavier's College
University of Bombay
Bombay, India

Jozef De Ley
Laboratory of Microbiology
and Microbial Genetics,
State University
Ghent, Belgium

Allen R. DeLong
Bellco Glass, Inc.
Vineland, New Jersey

Dick Detmar
Gist-Brocades N.V.
Delft, Netherlands

Shirley Dixson
Microbiol. Department
St. Vincent's Hospital
Fitzroy, Australia

Lars Edebo
Department of Medical
Microbiology
University of Linköping
Linköping, Sweden

Conny Eklund
Aminkemi AB
Bromma, Sweden

Eduard Engelbrecht
Department of Bacteriology
and Serology
Stichting Samenwerking
Delfise Ziekenhuisen
Delft, Netherlands

Eva Engvall
Wenner-Gren Institute
Stockholm, Sweden

Berthe Sigrid Erichsen
Forenede Margarinfabrikker
Oslo, Norway

Hans Ericsson
Department of Clinical
Microbiology
Karolinska Hospital
Stockholm, Sweden

Lens Ewetz
The Research Inst. of the
National Defense
Sundbyberg, Sweden

Akinyele Fabiyi
Virus Research Laboratory
University of Ibadan
Ibadan, Nigeria

Giovanni Fadda
Institute of Microbiology
University of Parma
Parma, Italy

Malcolm Falconer
Eastman Kodak Company
Health & Safety Laboratory
Rochester, New York

Robert Fildes
Glaxo Laboratories Ltd.
Fermentation Development
Department
Ulverston, Lancashire,
England

Walter Fisher
Difco Laboratories
Detroit, Michigan

Jud. B. Flato
Princeton Applied Research
Corporation
Princeton, New Jersey

Mohamed Aly Fouda
Department of Microbiological
Research
Egypt Ministry of Agriculture
Giza, Egypt

Ronald Freake
Ames Company
Division of Miles Laboratories
Elkhart, Indiana

Wilhelm Frederiksen
Statens Seruminstitut
Aalborg, Denmark

Leif Oddvar Fröholm
Methodology Department
National
Institute of Public Health
Oslo, Norway

Mack J. Fulwyler
Particle Technology, Inc.
Los Alamos, New Mexico

Daniel Y. C. Fung
Department of Microbiology
Pennsylvania State University
University Park, Pennsylvania

Rudolf Gallien
Hoffman-La Roche AG
Grenzach, West Germany

Donald M. Gibson
Ministry of Agriculture,
Fisheries and Food
Torry Research Station
Aberdeen, Scotland

Gordon Gifford
Beecham Research
Laboratories
Wothing, Sussex, England

Donald Glaser
University of California
Berkeley, California

Luigi Grassi
Laboratori Analisi
Ospedale Sesto
S. Giovanni
Milano, Italy

Rune Grubb
Institute for Med.
Microbiology
Lund, Sweden

Paul Groñnroos
Central Hospital
Tampere, Finland

Hans Gyllang
Pripp Bryggerierna AB
Bromma, Sweden

Helge G. Gyllenberg
Department of Microbiology
University of Helsinki
Helsinki, Finland

Bengt Gastrin
Bacteriology Laboratory
Central Hospital
Eskilstuna, Sweden

Gunther Hahn
Institute für Hygiene
Bundesanstalt für
Milchforschung
Kiel, West Germany

Lawrence B. Hall
Planetary Programs, NASA
Washington, D.C.

Walter Hameister
E. Merck
Medical Research
Darmstadt, West Germany

Brian Harrington
Mercy Hospital
Toledo, Ohio

Norman D. Harris
Department of Pharmacy
Chelsea College
London, England

Jim Harrison
UNESCO
Paris, France

Carl-Göran Hedén
Institute of Bacteriology
Karolinska Institute
Solna, Sweden

Walther Heeschen
Institut für Hygiene
Bundesanstalt für
Milchforschung
Kiel, West Germany

Karl Heil
Farbwerke Hoechst AG
Frankfurt/M.,
West Germany

Vera Heilborn
AB Kabi
Stockholm, Sweden

George Heimer
Public Health Laboratory
Central Middlesex Hospital
London, England

Robert E. Hinkel
U.S. Army
Frankfurt.M.,
West Germany

William Ian Hopkinson
Vickers Limited Research
Establishment
Ascot, Berkshire, England

Miles Hossom
Canalco, Inc.
Rockville, Maryland

Dennis G. Howell
Ontario Veterinary College
University of Guelph
Guelph. Ont., Canada

Tibor Illéni
Institute of Bacteriology
Karolinska Institutet
Stockholm, Sweden

Hal Ingram
Medical Technomics
Oakbrook, Illinois

Jan Isoz
LKB-Produkter AB
Bromma, Sweden

Adrian K. Jackson
Unilever Research
Laboratory
Sharnbrook, Bedford, England

Jindrich Janda
Oxoid Ltd.
London, England

Erik Jantzen
The State Institute for Public
Health
Oslo, Norway

Bundusiri Jayasekara
Yardleys of London
Basildon Essex, England

Moncef Jeddi
Bacteriology Laboratory
Hospital Charles Nicolle
Tunis, Tunisia

Gert Johansson
Köttforskningsinstitutet
Kävlinge, Sweden

Ingmar Juhlin
Bacteriological Institute
Malmö Hospital
Malmö, Sweden

Tage Justesen
Statens Seruminstitut
Copenhagen, Denmark

Lauri Jännes
Central Public Health
Laboratory
Helsinki, Finland

Lars Olof Kallings
National Bacteriological
Laboratory
Stockholm, Sweden

Martin Kaplan
WHO
Geneva, Switzerland

Karl-Axel Karlsson
The Research Institute of the
National Defense
Sundbyberg, Sweden

Gerald Kaufman
Perkin-Elmer Corporation
Wilton, Connecticut

Piet Kistemaker
FOM-Institute for Atomic
and Molecular Physics
Amsterdam, The Netherlands

Jan Kjellander
The Regional Hospital
Örebro, Sweden

Tapani Kohonen
National Board of Waters
Helsinki, Finland

Henry Kramer
Union Carbide Research
Institute
Tarrytown, New Jersey

Paul Donald Kronish
Department of Diagnostics
Research
Warner-Lambert
Research Institute
Morris Plains, New Jersey

Jozef Feliks Kubica
Wojskowy Institute for
Hygiene and Epidemics
Warsaw, Poland

Ronald Kundargi
Hospital of St. Raphael
New Haven, Connecticut

Johs. Kvittingen
Bacteriological Laboratory
Central Hospital
Trondheim, Norway

Emil Laga
Peter Bent Brigham Hospital
Boston, Massachusetts

Heinz Henning Lampe
Bayer AG
Leverkusen, West Germany

Stephen Lapage
National Collection of Type
Cultures
Central Public Health
Laboratory
London, England

Henri Lavirotte
Laboratoire B-D Mérieux
Marcy l'Etoile, France

Henri Leclerc
Institut Pasteur
Lille, France

John Leonardopoulos
Department of Microbiology
National University of Athens
Athens, Greece

John Lewis
Loma Linda University
Medical Center
Loma Linda, California

Owen M. Lidwell
Central Public Health
Laboratory
London, England

Arne Lithander
Serafimer Hospital
Stockholm, Sweden

Jorge Lopez-Trello
Antibioticos, S. A.
Madrid, Spain

Holger Lundback
National Bacteriological
Laboratory
Stockholm, Sweden

Arne Lundin
The Research Institute of the
National Defense
Sundbyberg, Sweden

Mackay-Scollay
Pathlab
Subiaco, Australia

Il-Young Maing
Bundesanstalt für
Fleischforschung
Kulmbach, West Germany

Ferruccio Mandler
Laboratoria Microbiologrea
Ospedale
Sacco-Vialba
Milan, Italy

B. G. Mansourian
WHO
Geneva, Switzerland

Pertti Markkanen
Helsinki, Finland

Arthur Markovits
Clinical Sciences Inc.
Whippany, New Jersey

Alexander Martens
Bausch & Lomb Inc.
Rochester, New York

I. A. Martin
Dept. of Pharmacy
London, England

Tom Martin
Allied Breweries
Burton-on-Trent, England

Michael McIllmurray
Wellcome Research
Laboratories
Beckenham, Kent, England

James E. McKie
Pfizer Inc.
Groton, Connecticut

Coenraad Meijer
Laboratory v.d.
Volksgezondheid
Leeuwarden, The Netherlands

Volker ter Meulen
Institut für Virologie
Universität Würzburg
Würzburg, West Germany

Henry Meunier
Sobioda
Seyssinet-Pariset, France

Henk Meuzelaar
FOM-Institute for Atomic
and Molecular Physics
Amsterdam, The Netherlands

Yvon Michel-Briand
Faculté de Médecine et
Pharmacie
Besancon, France

Dino Migliorini
Laboratoria Analisi.
Ospedale Fatebenefratelli
Milan, Italy

Brij Mitruka
Division of Health Science
Resources
Yale University
New Haven, Connecticut

Milton A. Mitz
NASA
Washington, D.C.

Nils Molin
Department of Applied
Microbiology
Karolinska Institutet
Stockholm, Sweden

Richard Moore
Central Public Health
Laboratory
London, England

Jaun-R. Mor
Institute of Microbiology
Swiss Federal Institute of
Technology
Zürich, Switzerland

Josephine Morello
University of Chicago
Chicago, Illinois

R. S. Moussa
Nestlé Products Technical
Assistance Co. Ltd.
La Tour de Peilz, Switzerland

Muscan
Biotest Serum Institut GmbH
Frankfurt/M, West Germany

Frédéric Müller
Institut de Physiologie
Lausanne Switzerland

Ulrich Müller
Battelle-Institut
Frankfurt/M., West Germany

Thomas R. Neblett
Department of Pathology
Henry Ford Hospital
Detroit, Michigan

Samuel W. B. Newsom
Department of Microbiology
Papworth Hospital
Cambridge, England

Slavko Neytcheff
Bulgarian Academy of
Medicine
Sofia, Bulgaria

Hans Aage Nielsen
Statens Seruminstitute
Copenhagen, Denmark

Tapio Niemelä
Department of Microbiology
University of Helsinki
Helsinki, Finland

Jorma Niemi
Department of Microbiology
University of Helsinki
Helsinki, Finland

Teruo Nishiyama
Sanko Junyaku Co. Ltd.
Tokyo, Japan

C.-E. Nord
National Bacteriological
Laboratory
Stockholm, Sweden

Staffan Norell
Coagulation Laboratory
Karolinska Hospital
Stockholm, Sweden

Erling Norrby
Department of Virology
Karolinska Institutet
Stockholm, Sweden

Urs Nydegger
Centre de Tranfusion
Hopital Cantonal
Geneva, Switzerland

Edwin Oden
Schering Corporation
Bloomfield, New Jersey

Tov Omland
Forsvarets mikrobiologiske
Laboratorium
Oslo, Norway

Joop Oostendorp
Gist-Brocades N. V.
Delft, The Netherlands

Örjan Ouchterlony
Bacteriological Institute
Sahlgrenska Hospital
Göteborg, Sweden

Francesco Parenti
Gruppo Lepetit
Milan, Italy

G. C. Parikh
Bacteriology Department
South Dakota State University
Brookings, South Dakota

Robin Pavillard
Department of Microbiology
Royal Perth Hospital
Perth, Australia

Holger Pedersen
Biocentralen
Hörsholm, Denmark

Olli Pensala
Technology Research Centre
of Finland
Food Research Laboratory
Ontaniemi, Finland

Kurt F. Petersen
Institut für Laboratoriums-
diagnese
Gauting, West Germany

Sven A. Petersson
N. V. Philips Gloeilampen-
fabrik
Eindhoven, The Netherlands

Michelangelo Petrini
Laboratoria Microbiologia
Ospedale Maggiore Ca`Granda
Milan, Italy

Jean-Daniel Piguet
Institute of Hygiene
Department of Bacteriology
Geneva, Switzerland

John Pike
Union Carbides Research
Institute
Tarrytown, New York

Michael A. Pisano
Department of Biology
St. John's University
Jamaica, New York

Gerard van-der Ploeg
Department of Medical
Microbiology
University of Nijmegen
Nijmegen, The Netherlands

Franco Porta
Laboratoria Analisi
Ospedale Civile di Sondrio
Sondrio, Italy

Julius Praglin
Pfizer Inc.
Groton, Connecticut

Vladimir Presecki
Department of Bacteriology
Institute of Public Health
of Croatia
Zagred, Yugoslavia

Franz Potsch
Bundesstaatliche Bakt.-Serol.
Untersuchungsanstalt
Vienna, Austria

Rudolf Quasi
Hygiene Institut der
Universität
Marburg, West Germany

Annikka Rantama
Orion Pharmaceutical
Helsinki, Finland

Fred Ray
Bio Quest
Cockeysville, Maryland

Jim Rechsteiner
National Institute of Public
Health
Bilthoven, Netherlands

Joseph D. Reilly
G. D. Searle & Co.
Chicago, Illinois

Nicola Riccardino
Laboratoria Analisi
Ospedale S. Giovanni
Battista Molinette
Torino, Italy

Egidio Rigoli
Instituto Microbiologia
e Immulogia
Ospedale di Treviso
Treviso, Italy

Ian Robertson
Adelaide, Australia

Åke Rosengren
Aminkemi AB
Bromma, Sweden

Manfred Rotter
Hygiene-Institutet of the
University
Vienna, Austria

Warren J. Russell
Instrumentation Laboratory
Lexington, Massachusetts

Per Ronsted
Novo Terapeutisk Laboratory
Bagsvaerd, Denmark

Alvar Sandkvist
AB Bofors, Nobel-Pharma
Mölndal, Sweden

Matti Sarvas
Central Public Health
Laboratory
Helsinki, Finland

Stanley Scher
New College of California
Sausalito, California

Werner Schneider
Uppsala Datacentral
Uppsala, Sweden

Elisabeth Schoutens
Hôpital Univ. Brugmann
Brussels, Belgium

Karlheinz Schroeder
Bad Oldesloe, West Germany

H.-R. Schulten
Institut für Physicalische
Chemie der Universität Bonn.
Bonn, West Germany

Hanfried Seyfarth
Dr. Karl Thomae GmbH
Biberach/Riss, West Germany

T. J. Sgouris
College of Medicine
Michigan State University
East Lansing, Michigan

Anthony N. Sharpe
Unilever Research
Laboratory
Sharnbrook, Bedford, England

David Shepherd
Nestlé Products Technical
Assistance Co. Ltd.
Orbe, Switzerland

Knud Siboni
Statens Seruminst.
Odense, Denmark

Poul Sigsgaard
Biocentralen
Hörsholm, Denmark

Carl Silfverstolpe
Silveca AB
Norsborg, Sweden

Libor Slechta
The Upjohn Company
Kalamazoo, Michigan

William R. Smith
G. D. Searle & Co.
Chicago, Illinois

Benkt Göran Snygg
Swedish Institute for Food
Preservation and Research
Göteborg, Sweden

Lennart Sparell
Swedish Water and Air
Pollution Research Laboratory
Stockholm, Sweden

Max Spinell
A/S N. Foss Electric
Hilleröd, Denmark

David Spooner
Microbiology Division
The Boots Company
Nottingham, England

Simone Stadtsbaeder
Cliniques St. Pierre
Université Louvain
Leuven, Belgium

L.-V. von Stedingk
Mikrobiol. Centrallab.
Stockholm, Sweden

Vilma Stenius
Orion Pharmaceutical
Helsinki, Finland

Charles E. D. Taylor
Public Health Laboratory
and Department of
Microbiology
Central Middlesex Hospital
London, England

Francoise Tchernia
Laboratoire microbiologie
Lavera, France

Ernst Thal
National Veterinary Institute
Stockholm, Sweden

Jean Thomas
Laboratoire d'Analyses
medicales de Wissembourg
Wissembourg, France

Sjur Thomassen
Apothekernes Laboratorium
Oslo, Norway

Dietrich Thon
Erprobungsstelle der
Bundeswehr
Munster/Oertze,
West Germany

Anders Thore
The Research Institute of the
National Defense
Sundbyberg, Sweden

Anthony Thorne
Dynatech AG
Zug, Switzerland

Adolf Tolle
Institut für Hygiene
Bundesanstalt für
Milchforschung
Kiel, West Germany

Robert Trotman
St. Mary's Hospital Medical
School
London, England

Kiyoshi Tsuji
The Upjohn Company
Kalamazoo, Michigan

Jerry Tulis
Becton, Dickinson &
Company Research Center
Research Triangle Park,
North Carolina

Nils-Erik Törnblom
Orion Pharmaceutical
Helsinki, Finland

Joseph Ungar
Liebenfeld, Switzerland

Amiram Ur
Clinical Research Centre
London, England

Richard Urbanezik
Schwarzwald-Sanatorium
Schömberg, West Germany

Bernard Vayssiere
Laboratory Lepetit
Suresnes, France

Da Cruz Ilda M. Vicente
Institute National de
Invest. Industrial
Lisboa, Portugal

Arturo Visconti
Laboratoria Micriobiologrea
Ospedale S. Carlo Borromeo
Milan, Italy

Torkel Wadström
National Bacteriological
Laboratory
Stockholm, Sweden

Francis A. Waldvoge
Clinique universitaire de
Medecine
Hôpital cantonal
Geneva, Switzerland

Clavin Ward
Cetus Scientific Laboratory
Inc.
Berkely, California

Motomi Watanabe
Sanko Junyaku Co., Ltd.
Tokyo, Japan

Peter Watins
Baird & Tatlock
London, England

A. H. Whaba
WHO
Copenhagen, Denmark

L. A. Williams
Kodak Ltd.
Harrow, England

Heinz Wolff
Clinical Research Centre
Harrow, England

Stanley Wolfsen
Clinical Sciences Inc.
Whippany, New Jersey

John Wood
British Food Manuf. Ind.
Research Ass.
Leatherhead, England

Bengt Wretlind
Department of Clinical
Microbiology
Karolinska Hospital
Stockholm, Sweden

Philip J. Wyatt
Science Spectrum, Inc.
Santa Barbara, California

Sven E. Young
S. E. Young Research
Laboratories Ltd.
Toronto, Ont., Canada

Eugene Yourassowsky
Hôpital Univ. Brugmann
Brussels, Belgium

K. Youssef
Misericordia Hospital
Medical Center
Bronx, New York

Klaus Yrjas
Oy Rohto AB
Ekenäs, Finland

Klaus-Jurgen Zaadhof
Lehrstuhl für Hyg.
und Tech. der Milch.
Univ. München
München, West Germany

Bengt Zacharias
The Research Institute of
The
National Defense
Sundbyberg, Sweden

Jadranka Zajc-Satler
Institute of Microbiology
Medical Faculty
Ljubljana, Yugoslavia

H. C. Zanen
Laboratory voor de
Gezondheidsleer
University Amsterdam
Amsterdam, He Netherlands

Vitali Zubov
Department of Chemistry
University of Moscow
Moscow, USSR

Rüknettin Ögütman
Medical School, Microbiology
Department Atatürk University
Erzurum, Turkey

Secreteriat

Sophie Olsen
Dept. of Appl. Microbiology
Karolinska Institutet
Stockholm, Sweden

Sylvia Molin
Swedish Nutrition Foundation
Uppsala. Sweden

Inger Öhman
Flygt AB
Solna. Sweden

SERIES PREFACE

This series of handbooks is an effort to supply the practical advice that is needed in most laboratories active in the various fields of applied microbiology and to do it without an overdose of theoretical considerations. This, however, does not imply that the books will be of only limited value to a theoretically oriented laboratory. Consider the extensive use of microorganisms as research tools now common among biophysicists, molecular biologists, immunologists, bioengineers and many others—and you will appreciate the need for a quick guide to the accepted techniques for handling bacteria and viruses.

Pure and applied microbiology go together; they are opposite sides of the same coin. The former is a road over forbiddingly steep hills on which the path is always partly hidden from view. The latter is the goal, for, after all, as Orville Wyss has emphasized, applied microbiology constitutes the backbone of our science, even if "we have responded to the gibes of the humanists who have always objected to the university leaving the cloister and entering the market place. It has never been demonstrated that the cloister is in any way superior to the market place for training a man to think, or that applied science is in any way inferior to pure science as an intellectual effort. "There are many signs that the young student generation is more keenly aware of this than most of their professors, but this should not make the students forget Louis Pasteur's famous statement: "Without theory, practice is but routine born of habit. Theory alone can bring forth and develop the spirit of inventions." If the student keeps this in mind he will find that microbiology offers more challenging opportunities to make inventions that will affect man's future health and well-being than most other subjects which he might choose to study.

Carl-Göran Hedén

PREFACE

The last couple of decades have witnessed a rapid sophistication in the methodologies employed in chemical laboratories, both in analytical work and in research. In microbiology, on the other hand, progress has been slow. It is now rapidly picking up speed, however, deriving inspiration from the immunological methods that serve as a bridge between the techniques used in biochemistry and those that require the propagation of cells. The latter group of course carries sterility constraints, and since the material studied may be pathogenic, it is natural that large-scale automated systems have been slow to appear. To some extent this has been compensated by the appearance of simple test kits suitable for handling large numbers of tests. To some extent this has generated a polarization between the advocates for automation and the promoters of simple kits.

In order to balance such opinions against each other and to provide an overview useful to United Nations agencies, hospital- administrators, and other persons responsible for the planning of laboratory services, an international symposium was organized. It was called "Rapid Methods and Automation in Microbiology" and took place in Stockholm between June 3 and June 9, 1973. It attracted some 310 specialists from 30 countries and was sponsored by UNESCO, WHO, and the International Organization for Biotechnology and Bioengineering (IOBB). It was supported also by grants from the Swedish Government, the Swedish Board for Technical Development and private industry.

Those papers that are concerned with new approaches to the identification of microorganisms are collected in this volume. Those that emphasize automation in microbiology and immunology appear in a second volume, *Automation in Microbiology and Immunology*.

The week-long conference also included a number of contributed papers, which we published in the abstract form (a list of them appears at the end of this volume). The discussions were lively—in fact, so lively that the original plans to publish them as part of the proceedings had to be abandoned. However, they have been prepared in a mimeographed version, which has been distributed to a large number of university libraries and which can also, like the abstracts, be obtained from the Department of Microbiology at the Karolinska Institutet at production cost ($5).

In the face of an exploding world population and growing demands for improved health services—despite lagging literacy in many parts of the world—the diagnostic productivity of the individual microbiologist must obviously be increased. Also, the opportunities that applied microbiology offers in such fields as fermentation, enzyme engineering, biological control and environmental management should constitute a challenge to many scientists who must now identify and screen innumerable potentially practical strains. Against this background we hope that this volume will have a catalytic effect.

We express our thanks to the symposia participants and to the various interested bodies for having made this book possible.

Carl-Göran Hedén
Tibor Illéni

Stockholm, Sweden
April 1974

THE SIGNIFICANCE OF IMPROVED MICROBIOLOGICAL TECHNIQUES TO THE UNITED NATIONS SYSTEMS

It may be of interest to mention one of the origins of this Symposium. In 1958 I began to attend, in a personal capacity, meetings of scientists from the East and the West to discuss problems of peace under the sponsorship of the Pugwash Conferences on Science and World Affairs. I was particularly concerned with problems of chemical and biological weapons and their possible use in international conflicts. Many Pugwash meetings were subsequently held on this subject and studies were made, assisted in part by the Stockholm International Peace Research Institute (SIPRI), which I believe helped greatly in arriving at the present international position of outlawing biological agents for warfare and the present negotiations on chemical weapons, the latter now pretty much at a standstill.

In February 1971 Pugwash convened a meeting of scientists on the rapid detection and identification of microbiological agents with a view toward assessing the technical feasibility of, first, community defense for a country exposed to actual attack and, second, the possibility of monitoring. The monitoring would provide some reasonable identification of open-air, clandestine trials—a likely step in the military development of biological agents —and the usefulness of such procedures in corroboration or refutation of allegations of use of such agents in a particular situation. (I might add that this was at a time when the problem of inspection and control was of particular importance in the negotiations for disarmament under way in Geneva. It still is a major obstacle in all disarmament negotiations.) At the same time it was believed that the techniques discussed during the 1971 meeting could contribute to the advancement of human health and well-being by application of some of the technological principles to medical and public health problems.

The Pugwash meeting of February 1971 was, in general, quite successful in achieving the purposes just stated. At WHO we were particularly interested in the potential of scientific and technological advances in improving the applications of knowledge available at that time, and the development of new procedures for diagnosis in epidemiological surveys and for monitoring environmental pollution. With the aid of consultants we subsequently designed a program to explore the diagnostic possibilities, and three of the men who assisted us in this respect are here with us today—Dr. Hedén, Dr. Taylor, and Dr. Mitz.

At WHO we have chosen to focus our attention for the next year or two on problems of major communicable diseases, including parasitic infections, which are of significance to large population groups throughout the world. We feel that much attention is already being devoted to the development of automated techniques relevant to individual clinical care, such as hospital laboratory and other individual diagnostic services. And that major problems in the poorer countries, such as tuberculosis, malaria, schistosomiasis, and cholera, tend to be neglected simply because they are not of great concern to the developed countries where advanced technological laboratories exist that are required for the high level of technological effort necessary to make progress.

As you will learn during the course of this Symposium, we have started activities in a few of these fields, but we are at the very early stages of development and there is long road to go.

All of you no doubt realize that the research and development required is a very expensive process in personnel and hardware, and the limited finances of WHO simply do not permit any appreciable laboratory activity of our own. Therefore, we depend on the goodwill and collaboration of research workers and institutes in both advanced and developing countries to achieve our purposes. This is the policy we have used so successfully during the past 20 years in WHO with respect to basic and applied research in many health and medical problems, including communicable diseases, nutrition, the biology of human reproduction, and mental disorders. The collaborative work we have stimulated is carried out through some 700 WHO designated or recognized international and national reference centers and collaborative institutes, to whom we sometimes give token financial grants whose effects are amplified by orders of magnitude through the use of their own already existing facilities. We very much hope that the participants in this Symposium will bear this fact in mind, and provide their help and collaboration to WHO. In this way we might together contribute our bit to lessening the misery of disease in many developing countries by improving our technological tools and thereby decreasing the time gap for applying corrective measures.

We appreciate the time and effort taken by Dr. Hedén to organize the present Symposium, which I am sure will carry us forward in the taks of improving our capability for medical and public health use of technological tools in the rapid detection and identification of microbiological agents.

M. KAPLAN
Director, Office for Science and Technology
WHO

CONTENTS

Part A New Technologies in the Automation of Microbiological Identification Routines

1   *Automation of Colony Identification and Mutant Selection*   3
     Donald A. Glaser

2   *The Modular Approach to the Automation of Microbiological Routines*   13
     Carl-Göran Hedén

3   *Cultivating Microorganisms on a Substrate Tape*   39
     H. Bürger and R. Quast

4   *Parallel-Multiple Analysis: A Promising Tool for the Study of Complex Networks of Biochemical Functions as Represented by Living Cells*   47
     R. Quast and H. Bürger

5   *Monitoring of Bacterial Activity by Impedance Measurements*   61
     Amiram Ur and D. F. J. Brown

6   *Rapid Automated Bacterial Identification by Impedance Measurement*   73
     Paxton Cady

7   *Bacterial Identification by Microcalorimetry*   101
     W. J. Russel, J. F. Zettler, G. C. Blanchard, and E. A. Boling

8   *Rapid Automated Identification of Microorganisms in Clinical Specimens by Gas Chromatography*   123
     Brij M. Mitruka

9   *High-Resolution Field Ionization and Field Desorption Mass Spectrometry of Pyrolysis Products of Complex Organic Materials*   155
     H.-R. Schulten

10   *Rapid and Automated Identification of Microorganisms by Curie Point
     Pyrolysis Techniques: I. Differentiation of Bacterial Strains by Fully
     Automated Curie Point Pyrolysis Gas-Liquid Chromatography*   165
     Henk L. C. Meuzelaar, Piet G. Kistemaker, and Anneke Tom

11   *Rapid and Automated Identification of Microorganisms by Curie Point
     Pyrolysis Techniques: II. Fast Identification of Microbiological Samples by
     Curie Point Pyrolysis Mass Spectrometry*   179
     Piet G. Kistemaker, Henk L. C. Meuzelaar, and Maarten A.
     Posthumus

12   *Identification and Grouping of Bacteria by Numerical Analysis of their
     Electrophoretic Protein Profiles*   193
     K. Kersters and J. De Ley

Part B Computer-Assisted Approaches to the Identification of Microorganisms

13   *Basic Principles in Computer-Assisted Identification of Microorganisms*   201
     H. G. Gyllenberg and T. K. Niemelä

14   *Problems and Approaches in Computer-Aided Analysis of Virus-Infected
     Cells*   225
     Hans M. Aus, Volker ter Meulen, Peter H. Bartels, and George
     L. Wied

Part C Advances in Epidemiological Surveillance

15   *Computer-Aided Monitoring, Surveillance, and Pattern Recognition*   255
     Richard Moore

16   *Experience Within a Computerized National Microbiological Data Retrieval
     and Analysis System*   275
     Thomas R. Neblett

17   *A Data System for Bacteriological Routine and Research*   291
     Stellan Bengtsson, Fred-Olof Bergqvist, and Werner Schneider

18   *Mobile Laboratory for Yellow Fever Studies*   307
     Akinyele Fabiyi

Part D   Current Trends in Simplified Diagnostic Tests

19   *New colony Markers due to Vital Staining During Growth on Dye
     Containing Agar*   317
     V. Bonifas, G. Demierre, and O. Ribeiro

20   *Enzymatic Analysis in Microbiology*   333
     H. Brunner and G. Holz

21   *Miniaturized Microbiological Techniques for Rapid Characterization of
     Bacteria*   347
     Daniel Y. C. Fung and Paul A. Hartman

22   *Development of Reagent-Impregnated Test Strips for Identification of
     Microorganisms*   371
     Donald P. Kronish

23   *Enterotube Roche: A Rapid and Accurate Method for the Identification of
     Enterobacteriaceae*   385
     Rudolf Gallien

24   *Evaluation of Different Diagnostic Kits for Enterobacteriaceae*   393
     Carl-Erik Nork, Torkel Wadström, and Ann Dahlbäck

25   *Evaluation of the API, the PathoTec, and the Improved Enterotube Systems
     for the Identification of Enterobacteriacea*   407
     R. S. Moussa

26   *Methods for the Rapid Identification of Enterobacteriaceae*   421
     J. Leonardopoulos and J. Papavassiliou

27   *Easy, Economic Typing of Enterobacteria*   435
     S. W. B. Newsom

28   *Method of Screening Cultures of Stools for Enteric Pathogens*   445
     Rüknettin Ögütman

29   *Primary Amoebic Meningoencephalitis: Agar Plate Method for Rapid
     Detection and Identification of Naegleria–Hartmannella Groups of Amoebae*   451
     S. R. Das

List of Abstracts   459

Index   461

NEW APPROACHES TO THE IDENTIFICATION OF MICROORGANISMS

NEW TECHNOLOGIES
IN THE AUTOMATION
OF MICROBIOLOGICAL
IDENTIFICATION ROUTINES

Automation of Colony Identification and Mutant Selection

DONALD A. GLASER

SUMMARY

Automatic equipment has been constructed for pouring agar into 40×80 cm glass trays, inoculating with microorganisms in a regular pattern, incubating the trays, making time-lapse series of photographs during incubation, analyzing the photographs with a flying spot scanner and computer combination, mechanically picking and reinoculating clones of interest, and presenting detailed information on the growth of each clone. The present semiautomatic prototype system has a capacity of 128 glass trays, and the fully automated system under construction will have a capacity of 256 trays able to carry 8×10^7 colonies at 1-mm spacing. The system is able to identify many bacterial pathogens by colony morphology and to find specific classes of mutants based on colony morphology, growth rates under controlled environmental conditions, response to drugs and nutrients, and other optically detectable traits. Current applications include evolutionary, genetic, and physiological studies of *E. coli* bacteria and will probably be extended to include studies of contamination, epidemiological monitoring, medical microbiological procedures, and transformation of animal cells in tissue culture.

INTRODUCTION

Many experiments on the molecular biology and genetics of bacteria call for the use of thousands of agar-filled petri dishes, involving much painstaking and monotonous work. These experiments require searches for rare mutants for the investigation of particular molecular mechanisms, measurements of spontaneous and induced mutation rates, genetic mapping of new mutants, and construction of exhaustive maps for organisms of particular interest. Similar laborious experiments are needed to study the rates and strategies observed in spontaneous evolution of microorganisms in the laboratory as well as to monitor the outcome of deliberate genetic engineering programs to produce strains with particularly desirable traits.

Among biomedical applications often requiring the use of large numbers of petri dishes are the following: monitoring the type, extent, and routes of contamination in food, water, air and other materials; studying the epidemiology of infectious disease; and carrying out procedures of clinical microbiology. With these and other applications in mind, we have been developing techniques and building equipment for automating many of the procedures required for growth of microorganisms on solid media. In the next sections we describe the methods and instrumentation for the automation, the means of collecting and analyzing the data, and a few preliminary results.

METHODS AND INSTRUMENTATION

Since the spontaneous mutation rate for many genes in *E. coli* bacteria is of the order of 10^{-8}, the materials-handling equipment was designed with a maximum capacity of about 10^8 colonies per batch, and data-handling equipment of matched capacity was provided. We decided to base all our measurements on optically detectable characteristics of growing colonies of microorganisms, concentrating our efforts on growth on solid media, where a great variety of growth conditions can be maintained and many colony characteristics can be measured. Optical methods lend themselves well to very high speed methods and to efficient and versatile use of computers.

In the final version of our machinery, agar will be poured into 40×80 cm glass trays, two trays being mounted on an aluminum frame of about 1 m^2. These frames are stacked in heat-sterilizable magazines containing 64 frames; the normal complement of each batch in the large machine, which is called the Dumbwaiter, is 128 frames. A moveable magazine, shown on the left in Fig. 1, carries a load of 128 glass trays which are heat sterilized and filled with agar by an automatic machine controlled by a small computer. After the agar has set, the moveable magazine is moved to the Dumbwaiter and mated with fixed magazines, as shown at the extreme left and the extreme right of the machine in Fig. 1. When both fixed magazines have been loaded with stacks of sterile agar-filled trays and sealed once again, the trays circulate in a clockwise direction, rising in the stack on the left, crossing the top tunnel toward the right, and moving downward through the stack on the right. The interior of the Dumbwaiter is a sterile incubator whose temperature will be maintainable in the range 4 to 50°C with a temperature accuracy of better than 0.1°C over most of that range. As the trays move through the top and bottom tunnels they will be accessible for photography; inoculation; addition of nutrients, drugs, and other chemicals; physical manipulation of colonies, and other necessary operations.

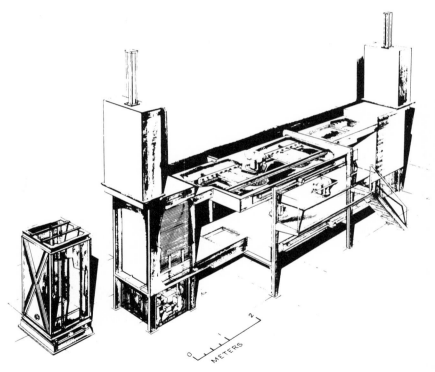

Fig. 1. Large-scale colony processor (the Dumbwaiter), used at the University of California Molecular Biology and Virus Laboratory for experiments involving up to 10^8 colonies of microorganisms. Colonies are carried on 256 agar-filled glass trays (40×80 cm) for automatic inoculation, incubation, photography, spraying with nutrients or drugs, mutant picking, and other computer-controlled manipulations.

At the beginning of an experiment, sterile agar is inoculated with organisms by a stream of fine droplets generated by a vibrating nozzle supplied with a suspension of organisms of suitable concentration. Both decks contain precision computer driven $x-y$ motions capable of positioning the trays with an accuracy and repeatability of about 25 μ over their whole surface. By means of this motion and using the vibrating nozzle inoculating system, all the trays in the Dumbwaiter are inoculated with organisms planted on a grid pattern with adjustable spacing from about 1 to 5 mm or larger, depending on the measurements to be made. The growth of colonies in regular rows and columns allows maximum packing of growing organisms without colony overlap; it also minimizes effects of contaminants not intro- duced through the inoculating fluid and materially reduces the computer time required for scanning and analyzing the data.

As the trays move through the Dumbwaiter at incubation temperature, they can be photographed at intervals; in addition, extra nutrients or drugs may be sprayed on the agar, portions of the growing arrays of colonies can be replicated onto fresh agar, especially interesting colonies can be picked for further study, and the growth temperature, gaseous environment, and light levels can be altered as desired. We now have in operation a prototype of this machine carrying out all the operations associated with one of the cross tunnels and a moveable magazine together with all the auxilliary warm rooms and cold rooms, the agar-making plant, and the computer data analysis system. The final machine is under construction and should be completed in 1975.

COLLECTION AND ANALYSIS OF DATA

Photographs of 10×10 cm^2 areas of agar are made using 35 mm black and white film carrying 24×24 mm images of each agar square. Generally, three photographs are made using a blue filter, a red filter, and a gray filter, to detect pigmentation, turbidity, and other color-sensitive properties of the growing colonies. A flying spot scanner on line with a computer is used to scan the 24×24 mm images at a resolution of 8000 lines, measuring 64 levels of gray on the photographic image. After the scanner has found a colony and determined its boundaries, a measurement of the optical density of the image is made at several hundred points across each of a few typical diameters of the colony; thus an optical density profile is generated for each colony photographed, using each of three different colors of light. A Fourier analysis is made of these profiles, and the diameters and heights of the major peaks and valleys of the profile are determined for a total of about 50 parameters for each colony (1).

A simple mutant hunt now underway requires that we isolate a number of mutants unable to synthesize DNA at 20°C but able to grow more or less normally at 37°C. Newly inoculated colonies are incubated for a short time at 37°C after mutagenesis, giving mutants a chance to express their phenotypes. By dropping the temperature to 20°C for a period of time, then raising it again to 37°C, it is possible to identify among the survivors cold-sensitive mutants—namely, those which grew slowly or not at all at the restrictive temperature. This method is comparable in its yield with the conventional, classical method of replica plating, but it can be carried out completely automatically.

Our preliminary experiments involving only about 40,000 colonies in semiautomatic batch mutant hunts show a labor and agar saving of a factor of 5 compared with manual replica-plating methods. For larger scale automatic operation, the labor requirement increases very little, although the labor requirement for manual operation increases linearly with the number of colonies examined. It is hard to compare manual and automatic methods for colony diameter and colony morphology mutant hunts, since only the scanner is able to measure quantitatively the required morphological parameters. It is impractical to carry out such measurements without the machines.

For contamination analysis the computer must produce a more sophisticated identification of bacterial species and strains, which it does by making a cluster analysis of the 50-parameter description of the colonies (1). In effect the computer is shown about 1000 examples of each of the organisms expected to be found in a contamination sample, from which it "learns" the typical values of the parameters to be expected and the error distribution of these values. When presented with an unknown colony, the machine assigns a probability that the colony belongs to a particular species of interest. By setting probability acceptance thresholds for this identification, the operation can minimize the rate of incorrect identifications. In a test of this kind involving eight species of bacteria, the identification of an unknown colony into one of eight previously characterized species was done with an accuracy of better than 92% for most species (1). Improvements in the program have raised the accuracy to about 98%.

We are presently working on a program that will analyze a larger library of "standardized" organisms. The limitations of the method are partly limitations in the size and speed of the computer. Our computer system now requires about 20 sec to carry out this complete analysis, involving the scanning, data analysis, and species identification for three photographs of a standard 10-cm round petri dish containing about 50 colonies. It is easy to imagine reducing that time to 10 sec or perhaps a little less.

Another limitation is the need for great uniformity of environmental conditions and growth medium to assure reproducible colony morphology. Therefore, rather elaborate designs have been developed for careful incubation, photography, inoculation, and preparation of agar. A third limitation is that the microbiological sample must be preprocessed so that it finds itself in the standard environment when allowed to grow for the colony morphology test. This does not pose a serious problem for contamination monitoring and epidemiological studies, but for medical diagnostic applications a fair amount of development work will probably be necessary for preparation of medical specimens in a standard form.

Most of our colony morphology identifications have been done on colonies growing for 24 hours, but that time might be shortened somewhat depending on the organism. Once the computer has identified a particular colony, the machine will be able to retrieve it physically or to administer an antibiotic to see whether the organism can be killed or inhibited by administration of that antibiotic. Since typical colonies in our tests grow at the rate of several microns per minute, the expected diameter change, if the growth is not inhibited by the drug, is about 0.1 mm/hour, which is readily detected in our system. It should therefore be possible to read the results of drug-sensitivity tests within about an hour after the organism has been identified.

APPLICATIONS

A design prototype of the Dumbwaiter (called Cyclops because it has only one camera) is now being used for mutant hunts and will soon be used for mapping mutants and for making exhaustive genetic maps of *E. coli*. At the same time we are developing larger and more rapid programs for recognizing colony morphology types to be applied to contamination and epidemiological problems and perhaps in the future to problems of clinical microbiology. We want to use the machines also for producing particular genotypes needed in our experiments and for studying the rates of gene duplications, deletions, translocations, point mutations, frameshifts, and other kinds of chromosomal changes in configuration and size that play a role in bacterial evolution.

We plan to add a television viewing station to the machine, which will permit us to perform some experiments in real time. The computer will then be instructed to intervene in the experiment while it is happening, according to criteria specified in advance. For example, the machine might be instructed to change the temperature when colonies of standard morphology recognizable by the machine have reached a certain size. The machine will be controlled by a small computer capable of monitoring the temperature at numerous points as well as operating the cameras, lights, colony manipulators, and other accessories to the machine.

The analysis of data with the flying spot scanner is done by a larger computer that produces magnetic tapes containing instructions for further manipulations by the small computer. In this way the scanner finds addresses of mutants of interest and decides where those colonies should be put for the next stage in the operation. The small computer then carries out these operations using the instruction tape from the large one. In the future direct communication between the computers will make the physical transfer of magnetic tape unnecessary.

There is nothing that intrinsically limits the use of the machine to the growth of bacteria. In fact, the environmental control system has been designed with some flexibility to accomodate other microorganisms and animal cells growing in tissue culture. In effect, the Dumbwaiter is a machine for presenting about 82 m^2 of very closely controlled biological growth surface to periodic examination by a computer, using a flying spot scanner and photographic film or a television camera directly. It is able mechanically to move this entire area of growth surface past an inspection station in about 20 minutes, although it requires about an hour and a half if a photograph is taken of each 10×10 cm square.

Acknowledgments

Major contributions to the design and construction of the machine were made by R. Baker, J. Berk, T. Fujita, L. Hansen, R. Henry, A. Para, and P. V. Peterson; to the construction, programming, and operation of the data analysis system by J. Berk, F. Bonnell, J. Couch, R. Henry, and C. B. Ward; and to the carrying out of biological experiments using the system by J. Couch, J. Raymond, P. Spielman, C. B. Ward, and C. T. Wehr. The main support has come from the U.S. Public Health Service through research grants GM13244 and GM12524 from the National Institute of General Medical Sciences, with additional support from the University of California, the IBM Corporation, and NASA.

REFERENCES

1. D. A. Glaser and C. B. Ward, Computer Identification of Bacteria by Colony Morphology, in *Frontiers of Pattern Recognition*, Academic Press, New York, 1972.

CHAPTER 2

The Modular Approach to the
Automation of the Microbiological Routines

CARL-GÖRAN HEDÉN

SUMMARY

Early experience with cultivation on moving agar tapes indicated an interesting potential for identification by optical scanning but also limitations in the choice of media and gas phases. To achieve a greater flexibility, the "Autoline" system was developed. Its central module is a cutting device that produces 480×10 mm long and 2 to 5 mm thick gel ribbons, which are deposited on glass strips or sandwiched on gels already deposited on glass strips. Other modules provide for easy application of up to 48 different diffusion centers (punched holes, filter paper, etc.) per strip (antibiotics, sera, phages, etc) and for the rapid and automatic handling of large numbers of samples. The usefulness of long concentration gradients is discussed, as well as the potential of the system for resistance testing, identification work, immunodiffusion studies, and air, water, and surface sampling. The strips are well suited for simple optical scanning and for pattern recognition.

INTRODUCTION

Microbiology suffers from the lack of a methodological mean denominator capable of stimulating the development of mechanized approaches that would pave the way for automation making full use of the opportunities now offered by computer science. This Chapter emphasizes that gel techniques might be such a mean denominator, as indicated by the innumerable modifications of solid media cultivation, immunodiffusion techniques, and various zymographic and electrophoretic methods, which illustrate that the phenomena on or in gels are remarkably flexible. Taken with the growing number of well-defined polymers that are now available and the rapid development of pattern recognition techniques, they indicate that there is a growing need for mechanization of gel handling and analysis.

Diffusion in gels can introduce a simple dilution parameter in many techniques, and this eliminates the need for discontinuous or continuous measuring and mixing operations. True enough, diffusion takes time; but in many cases it may be finished before the actual test starts, or at least within the time required for growth or for some enzymatic reaction to run its course. With regard to optical scanning, the gel systems also offer specific advantages. For example, particulate reactants (cells or particles) can be located at a defined plane or along a fixed line either at a gas/gel interphase or inside a gel sandwich made up of identical or dissimilar polymers. This permits easy supply and removal of low molecular weight substances and increases the opportunities for cell or particle interactions. In fact, the potential of the interphase for separating molecules and cells is probably very far from being exhausted methodologically.

Nearly a decade ago, in 1965, when Falch and I developed a "band machine" for automated selection of lysine-producing strains (1), we pointed out that similar techniques could be used for identification purposes:

> The method might be extended to develop an automated device for identifi-cation of microorganisms. In this device, a series of parallel agar bands of different composition would be inoculated simultaneously with the same organism. The bands would be read by a photometric detector and the information processed by an analog conventor and digital computer or by a Memistor device.

Indeed, NASA later followed such an approach in one of their experimental designs for the manned space laboratory. However, our original device was based on feeding melted agar on to a moving polypropylene tape, and the quality of the medium could not be guaranteed over long periods of operation. Even worse, switching media was difficult, and individual samples could not be provided with their special gas mixtures. From one obvious solution—namely, cutting the tape in shorter pieces, one per sample—came a number of laboratory prototypes which we called "Autoline" units, to emphasize the automation goal and the principle of linear scanning (2, 3).

THE BASIC CONCEPTS OF THE AUTOLINE SYSTEM

The needs for mechanization and automation in microbiology are considerable, but microbiology is still far from reaching the degree of development attained by clinical chemistry. There are several reasons for this—for instance, the special problems involved in the handling of pathogens. However, the major obstacle to automation has been the very wide spectrum of methods utilized in microbiological work. This has generated the inertia that prevents rapid acceptance of many of the sophisticated physicochemical techniques now available. Some of them permit a remarkable saving in time and labor, but the narrowness of their impact sectors makes it hard for hospital administrators responsible for small institutions (up to 200 beds) to accept the high capital or operating costs that are often involved. Against this background, I felt that a highly flexible, modular approach was needed.

A modular approach to automation in microbiology should accommodate a large number of established routines and should provide such a convincing perspective toward incorporating the original modules into a fully automated system that any limitations in their performance—when compared with the specialized techniques—would be fully offset. I also felt that there was scope for designing core modules in which the expensive mechanical and optical parts were concentrated. Those core modules should then accommodate subunits geared to the specific needs of individual scientists or routines. The Autoline system is the result of such considerations. Let us now consider the two core modules that perform the basic functions outlined in Fig. 1—namely steps I, II and A, B in Fig. 2. Those functions are actually performed by two boxlike devices: the agar strip cutter and the apparatus for applying samples and diffusion centers.

An ordinary petri dish (9-cm diam.) contains about 25 ml of a nutrient agar medium with a depth of 0.33 cm. If this were instead spread out in the form of a 1-cm straight ribbon of the same thickness, it would be about 75 cm long. However, a linear surface could be better utilized than a round one, so I settled for a 0.5-m tube as our basic cultivation unit (Fig. 3). It contains a 1.5×50 cm carrier strip for an agar ribbon and is normally closed at one end with an injection flask rubber stopper through which gas can be introduced or removed for sampling. In the early work the other end was simply closed by an ordinary rubber stopper. As compared with the petri dish, the glass strip only carries two-thirds of the nutrient agar in a standard petri dish; further economizing is possible by some reduction in thickness and by using a sandwich technique for certain compounds, such as blood.

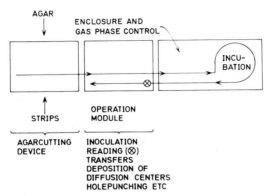

Fig. 1. Functional relation between the Autoline "core" modules.

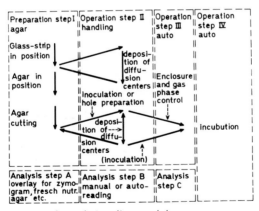

Fig. 2. Transport sequence through Autoline modules.

In the first core module (Fig. 4) various types of solid media blocks can be introduced easily. They are then subdivided into accurately shaped gel strips, which are deposited on the transparent carriers, normally made of glass (Fig. 5).

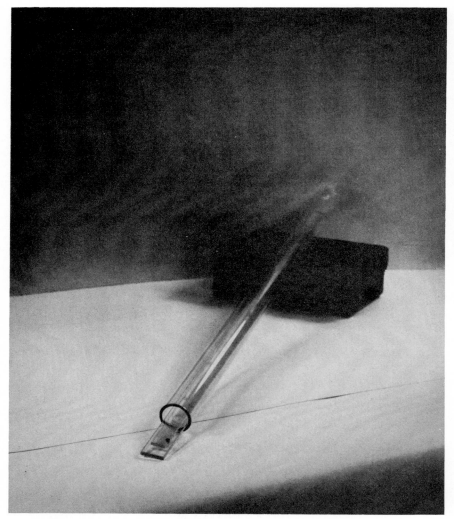

Fig. 3. Glass carrier with agar strip in tube.

In the second core module (Fig. 6) various groups of diffusion centers can then be applied. One scientist might want to use a complement of phages, another a group of antisera, and a third might want to perform a taxonomic study on the basis of a collection of carbon sources, inhibitors, and enzyme substrates.

Supplementary modules, taking care of such problems as gas phase

Fig. 4. Unit for cutting agar blocks.

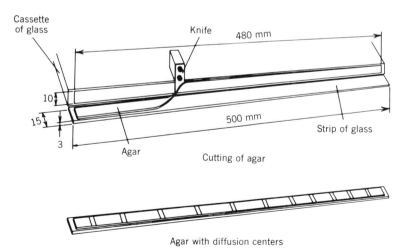

Cassette of glass

Knife

480 mm

10

15

3

Agar

500 mm

Strip of glass

Cutting of agar

Agar with diffusion centers

Fig. 5. Production of agar strips.

control and analysis, automation of the incubation time, etc., are being developed; thus a comprehensive system will eventually emerge (Fig. 7). My goal is to make this system so compact that it will be able to function in a mobile laboratory requiring only one microbiologist and a technical assistant for a considerable throughput. The road to this goal will be a long one, and we still work with a very crude system for handling the large number of tubes that the system generates (Fig. 8). However, I visualize no major hurdles to the development, besides the problems of funding, which most of us face.

Fig. 6. Unit for sample application and deposition of diffusion bands.

SPECIAL ATTRACTIONS OF THE AGAR STRIP TECHNIQUE

The main attraction of the agar strip technique was the possibility of developing simple optical scanning systems easily adaptable to the self-scanning photodiode chips that are now becoming commercially available. There may be a considerable potential need for such systems in microbiology

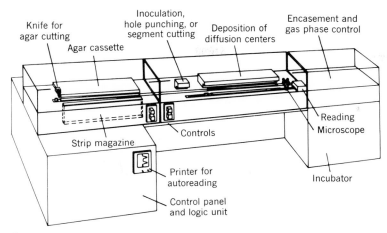

Fig. 7. Layout of a comprehensive Autoline system.

(Table 1), which can also learn much from the scanning systems used in chemistry (4–6).

However, since this aspect is discussed elsewhere (7) I will not describe our efforts to develop a flexible optical system for the Autoline. Rather, I want to emphasize that the ease of scanning was only one of many arguments for an agar strip system (Table 2). For instance, the possibility of slicing up thin agar blocks allows us to economize the medium. This may be of considerable importance both in cultivations involving expensive chemicals and in gel diffusion work requiring exotic reagents, e.g., in the determination of influenza antibodies by virus-containing gels (8). Considering its potential practical significance for influenza antibody surveys, complex gel diffusion work is one of the areas in which we have proved the value of our optical scanning system (Fig. 9), (9).

Another very attractive feature of the agar strip technique was the inherent possibility of standardizing all environmental parameters to extract more taxonomic information from the colony morphology, the metabolism, and the antigenic structure of microorganisms. Important in this connection is the accurate determination of the gel thickness ($<5\%$ error); in addition every sample can be offered a fresh surface. It is in fact likely that aging of plates, involving dessication and oxidation of the surface, is not only important for substrates like the lecithin agar plates used in the cultivation of anaerobes, but also in everyday diagonostic work, where the differences may, however, be masked by the effects of the normally tolerated variations in the incubation time.

Fig. 8. Experimental rig for automatic handling of large number of agar strips.

PREPARATION AND TRANSPORT OF AGAR STRIPS

The gel strips are cut from 5 or 10 mm thick blocks normally prepared from nutrient agar. They are pushed out from horizontal casettes containing from 200 to 1000 ml of solid medium, depending on their size. The strips are then deposited on a sterile glass or plastic carriers appearing from a storage box, located along the front of the cutting machine. The carrier strip is positioned on a transport chain that will move it along to the right into the next core module, to which it travels under a constant sterile air overpressure. In this

Table 1 Capabilities of Current Scanning Techniques

Need	Growth Film Scanning	Colonies Developing Along a Line	Colonies Developing on a Uniform Surface	Colonies Developing on a Gradient Surface	Reactions of Particulate Enzyme Substrates or Indicators	Flourescence of Nonflourescent Substrates	Flourescence of Labelled Sera	Luminescence	Precipitates
Identification and numerical taxonomy	×	×			×	×	×		×
Resistance testing	×			×	×			×	
Microbiological analysis	×				×			×	
Metabolic health screening	×					×			
Auxotyping	×								
Phage typing	×								
Selection work				×					
Surface sampling			×						
Air and water sampling			×						
Sterility and viability determinations		×	×						
Serological studies							×		×
Immunoelectrophoresis						×			×

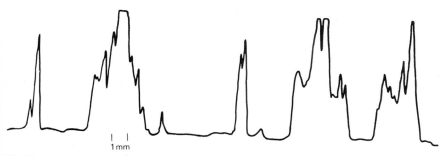

Fig. 9. Optical scan of reactions between gel containing influenza virus and sera in 1-mm φ wells.

Table 2 The Agar Strip as an Alternative to the Petri Dish

1. Geometry invites mechanization and permits sandwiching, application of re-agent tapes, and other modifications.
2. Wide choice of materials for simple and cheap manufacture of tubes and carrier strips.
3. Closure and gas phase sampling easy.
4. Good sterility control and anaerobic transfers possible.
5. Good heat transfer in spite of high packing density of tubes.
6. Systems dishwashing easy.
7. Moist, unoxidized cultivation surface provided.
8. Medium economy good, and supply (agar thickness) easily controlled.
9. Application of diffusion centers easy.
10. Inoculation and sample application simplified.
11. Long concentration gradients easily generated.
12. Adaptable to automated reading:
    a. good optical characteristics of surfaces involved
    b. illumination through side of agar possible
    c. determination of positions easy
13. Adaptable to sequential air and water testing and to the sampling of uneven surfaces.
14. Adaptable to gel permeation chromatography, electrophoresis, and other laboratory practices.

"operation module" diffusion centers are deposited, and the inoculation can be performed. During the transport from left to right, a punching device, located in the operation module, can be activated, providing the agar strip with a sequence of diffusion wells and/or dividing it into segments (Fig. 10).

INOCULATION AND SAMPLE APPLICATION

For the performance of resistance tests, microbiological analyses, and meta-bolic pattern studies, the quality of the bacterial film that is spread over the surface is critical. A disposable porous ceramic roller held stationary in relation to the moving agar strip has been found to provide the optimal solution (Fig. 11). In critical studies, it might be advisable to standardize the cell suspension nephelometrically.

Inoculations for colony isolations are simply performed by moving the platinum loop back and forth across the moving agar strip, This produces a zigzag line that is very long and, consequently, effects a considerable dilution.

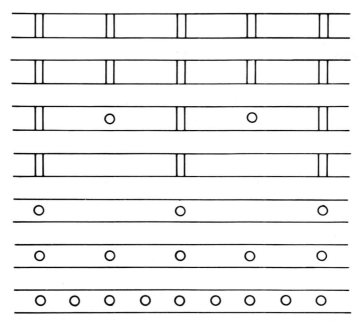

Fig. 10. Various possible combinations of diffusion bands and wells.

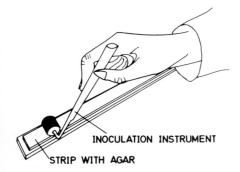

INOCULATION INSTRUMENT

STRIP WITH AGAR   **Fig. 11.** Inoculation roller.

Finally, with regard to the deposition of microorganisms in such a way that single colonies develop on the centerline of the strip, an electronic gate mechanism for single cells would probably be ideal. However, we have found that point inoculations with a 0.025-mm platinum wire will also yield colonies suitable for making an "optical cross section." The major problem with this technique is the risk of causing minute agar wounds that are invisible to the unaided eye but might be detectable by the photometer.

Liquid samples are handled by ordinary capillary pipettes, but a special dosing device is under development.

THE HANDLING OF DIFFUSION CENTERS

The dimensions of the agar strip make for simple mechanization of the application of diffusion centers. They are supplied in the form of 2-mm-wide bands coming off 48 reels turning around an axis parallel to the agar strip. In the course of being simultaneously cut off, many parallel pieces 1 cm long from selected reels can thus be deposited either directly on the glass strip (i.e., under the agar strip later provided) or on the surface of the solid medium.

Inhibitors and growth factors are normally carried by filter paper bands, but gelatin would be preferred to paper, not only for solid enzyme substrates (starch, lipids, elastin, DNA, RNA, red blood cells, etc.) but also for sera and bacteriophages. Gelatin has the advantage that it melts and effectively releases its contents at an incubation temperature of 37°C. But it can also be hardened by formalin to prevent melting. This is done, for example, if one wants to detect gelatinase by using a gel containing carbon particles (10). If the gel is hardened, the carbon particles will only escape and float out on the agar surface under the influence of enzyme action.

To detect other proteases, the precipitation of casein would seem to be most appropriate (11), possibly followed by $HClO_4$-treatment, which shows when the hydrolysis reaches a point at which casein can no longer be precipitated (12).

SOME FIELDS OF APPLICATION

Antibiotics Resistance Testing

In recent years the increased attention to resistance development in bacteria has emphasized the need for sensitivity testing in preparation for therapy. Particularly serious is the appearance of coupled resistance (e.g., to ampicillin, streptomycin, choramphenicol, and sulfa) as a consequence of tetracycline treatment of urinary tract infections (13). This type of resistance, which is mediated by R factors, points to the dangers of a broad-spectrum approach and underscores the need for a therapy geared specifically to the etiological agent. Against this background, the great attention now given to the developemnt of new methods for cheap and rapid resistance measurements on a large scale are very important. They will permit the controlled therapy now needed when we have finally learned the lessons from the three earlier phases of therapeutic practice: the switching back and forth between narrow-spectrum antibiotics, followed by the period of broad-spectrum drugs, and finally the period of combination therapies.

Since the use of the agar strip method for antibiotics resistance testing has been described by Dr. Wretlind (14), let me simply note again the advantages of limiting the inhibition zone measurement to one dimension. Not only does this permit the use of simple scanning and printout devices, but it also allows for a substantial reduction in the substrate requirement. However, some other aspects should also be mentioned.

The simple deposition of large numbers of diffusion bands or disks in various configurations can be used to disclose synergistic and antagonistic effects between various antibiotics. As an example, Fig. 12 illustrates how the diffusion centers for some antibiotics would be arranged to provide zone overlaps that could disclose synergistic effects that might be important in the treatment of urinary tract infections.

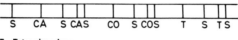

S   CA   S CAS   CO   S COS   T   S   TS

T = Trimetoprime
S = Sulfametoxazole
CA = Carbenicilline
CO = Colistinmethansulfonate

Fig. 12. Example of uneven distribution of diffusion bands aimed at achieving zone overlaps, illustrating synergistic effects.

The agar strip method is also especially well suited for bioautography of antibiotics based on paper strip chromatograms brought into contact with a film of sensitive microorganisms. The resulting R_f values can then be used to characterize the substances. It should be noted that this technique can also be applied to other biologically active compounds (e.g., antitumor, antiviral, or antifertility agents, tranquilizers, and cholesterol-lowering compounds), which often have antimicrobial properties as well (15).

A possibility of even greater practical significance is the establishment of very long concentration gradients based on a single antibiotic contained in filter paper bands, charged with increasing quantities from one end of the agar strip to the other. If the filter paper bands are located under the agar, the surface of the medium can be inoculated with a suspension of bacteria that will develop only up to a certain point on the gradient. Here the growth limit may be quite sharp, as indicated by Fig. 13. Of practical significance is the fact that the agar surface can be inoculated with a mixture of microorganisms. When these have developed into colonies, a preliminary morphological diagnosis can be made while the resistance level is being estimated on the basis of the height on the gradient that can be reached by the various colonies. As far as I know, this is the only possibility so far available for cutting the time required for resistance measurement down to zero, once isolation has been achieved. Of course the technique has the added advantage that the response becomes independent of the diffusion speed (molecular weight) of the antibiotic tested.

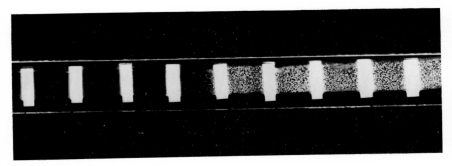

Fig. 13. Appearance of zone border on 48-cm-long gradient.

With regard to the determination of antibiotics in body fluids, the use of wells in the agar strip seems to be the method of choice. It may well be combined with various acceleration methods, ranging from the use of metabolic indicators to the luciferin/luciferase or luminol methods. In fact, our scanning system is now being adopted for the latter type of modification which, however, also requires a simple device for exposing the moving agar strip to a transparent reagent tape, thus permitting the measurement to be synchronized with the maximum photone yield. This effort will be given high priority because of the growing importance of rapid determinations of the serum levels of such antibiotics as gentamycine.

Microbiological Analysis

The use of a medium that lacks a growth factor (e.g., an amino acid or a vitamin), together with a bacterial strain that has the corresponding growth requirement, permits the quantitative determination of the substance in various samples (16). Using surface inoculation, the method has been successfully applied to lysine, tryptophane, threonine, and methionine down to 0.01 μg per diffusion center. Its potential for screening metabolic disorders such as phenolketonuria is obvious.

Auxotyping

Various bacteria and even individual strains may exhibit characteristic nutrient requirements and metabolite patterns. Such patterns may establish a bacteriological diagnosis, as in the case of Enterobacteriaceae, for which a number of commercial kits are already available; or they may have a great potential for epidemiological studies, as in the case of gonococci. The latter can not be successfully typed by phages, bacteriocins, or serological reactions, but Carifo and Wesley-Catlin have recently demonstrated 28 auxotypes with distinct requirements for proline, arginine, serine, methionine, isoleucine, hypoxanthine, uracil, thiamine, and thiamine pyrophosphate (17).

The taxonomic relevance of the volatile metabolites has now been so clearly illustrated in the case of anaerobes that a flexible cultivation system must be easily adaptable to gas sampling. This is a consideration that influenced the "Autoline" design at an early stage. However, even more important were the requirements for gel scanning that emerged from the effort launched jointly by Prof. H. Gyllenberg and his group in Helsinki and my group in Stockholm. The strategy behind our project, which focused on numerical taxonomy, is outlined elsewhere (18). Thus I need say only a few words about the required hardware, which had to permit the testing of a microorganism on a very large number of different solid media. Relatively high concentrations of critical nutrients in a base medium are then needed, and diffusion zone overlaps must be avoided.

Both the foregoing requirements could be met by cutting the agar strip into segments (Fig. 14). This can be done automatically—with a rectangular punch, cutting right across the agar strip. Since the smallest transport step for the agar strip is 10 mm, the strip must be moved 5 mm to the right or left before the diffusion bands are deposited. If the reaction is to proceed in a well, as when one wants to catch a gas bubble under a drop of mineral oil, two diffusion bands loaded with the same nutrient would be used on each side of the well. If, on the other hand, only small blocks are to be charged with the nutrient and suitable indicator, one diffusion band per block would be adequate. Figure 15 shows the appearance of a test with *Klebsiella pneumoniae* on a medium incorporating phenol red or bromthymol blue and charged with a number of sugars. In Fig. 16, on the other hand, we see a photometric signal pattern from a test with enterococci, illustrating where an electronic cutoff level would be applied in the course of computer processing.

Microbial Genetics

Five types of mutants are of special practical significance. *First*, there is the classical zone mutant where, for instance, extra-large zones

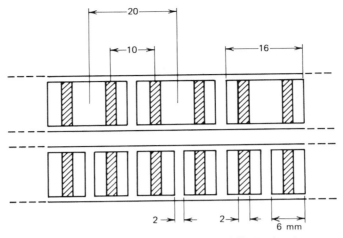

Fig. 14. Segmented agar strips charged from diffusion bands.

Fig. 15. Appearance of *Klebsiella pneumoniae* on various sugars (glucose, sucrose, sorbitol, adonitol, and galactose + phenol red and finally various concentrations of xylose/and urea + bromthymol blue).

around cephalosporin-producing fungi are selected (19). *Second*, there is the culture which needs an inducer to produce a desired enzyme, (e.g., β-galactosidase) and where one looks for large zones around constitutive mutants grown on agar without inducer (20). *Third*, one occasionally needs a mutant where the production of an enzyme, like invertase in Saccharomyces, is resistant to catabolite repression. Colonies are then selected from a medium rich in the repressive carbon source (21). *Fourth*, one might be looking for mutants which are resistant to product inhibition. Higher streptomycin yields have for instance been obtained by selecting colonies from mutagenized populations spread on gradient plates (22). And *finally*, one might want to produce cells which are resistant to an adverse pH or an elevated temperature.

In all those situations, particularly in the case of product inhibition and temperature mutants, long gradient strips might be very useful, and optical scanning might be helpful considering the large number of colonies that must be screened.

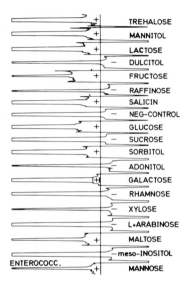

Fig. 16. Photoelectric scan over enterococci growing on agar segments with various sugars.

Serology

Two major application areas deserve special attention. One is the Manzini technique of using antibody-containing gels to determine the protein composition in serum samples. Optical scanning and printout of the results may be of great value in this area.

We are also going to consider the practical use of the optical phenomena that appear in microbial films exposed to homologous antibodies. Here the "Autoline" system is significant because a bacterial film can be exposed to a large number of sera simultaneously. In certain cases we have noted that our optical system picks up the reactions with homologous serum quite easily. In other cases one might have to fall back on fluorescent antibody techniques to detect a reaction. Here the air/gel interphase might be used to eliminate the need for the extra separation steps often involved in the indirect technique. Suppose, for instance, that a film of microorganisms is exposed to a large number of rabbit sera separated in space on a gel made strongly absorbing for the light used to exite a fluorescent antirabbit serum to be added later. Given sufficient time for diffusion down through the gel, only the homologous serum would remain on the surface, where it could be reacted with the fluorescent antirabbit serum when this was added. If epiillumination was then used after a certain delay, only the area reacted with homologous serum would emit light. In principle; such a system would permit simple testing of a bacterial growth film against a very large number of sera.

Finally, it goes without saying that the possibility of making wells in the gel strip should stimulate the development of immunodiffusion applications in the presence or absence of an electric field.

The well punch might be operated along three lines: in the middle of the agar strip or along one of its sides. By running the punching process several times, a well geometry suitable for Ouchterlony tests can be established. In many such applications, the optical scanning and the opportunities for computer analysis of the results should hold great attraction.

Surface Sampling

It was early noted that the geometry of the agar strips would not only be relevant to air and water sampling, but that it might also offer special advantages in surface sampling in that long cross sections of the distribution of microorganisms might be obtained. Indeed a grid of linear sampling surfaces might present entirely new opportunities for making statistical studies of the populations on uneven surfaces like the human body. Unfortunately, the agar strips are rather brittle, and cotton bands or other similar supports had to be used. However, the adhesion to the agar surface was not always adequate, and the ideal solution was not found until we tested 10-mm-wide nylon bands covered with minute hooks (one element in the band pair used by the clothing industry as an alternative to zippers). The hooks could easily be forced into the back of the agar strip, which was thus provided with a "handle" that made its use very simple (Fig. 17). Studies revealing the potential of the agar-strip technique will be published eleswhere (C.-G. Hedén and P. G. Hedén), and it is illustrated here only with two pictures of an open sandwich left overnight at room temperature: Fig. 18 is a direct photograph and Fig. 19 the "bacteriogram." The latter may be difficult to analyze in the picture, but the direct study of the solid substrate clearly brought out the preferential location of various microorganisms (*Pseudomonas*, *Enterobacteriaceae*, and gram-positive cocci).

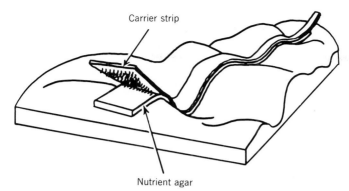

Fig. 17. Sampling from uneven surface.

Fig. 18. Open sandwich left overnight at room temperature.

FUTURE DEVELOPMENTS

From a methodological point of view the opportunities for simple sandwiching techniques may prove to be particularly useful, and various types of enzymograms and toxinograms can be visualized. It is also likely that the development of special types of agarose will generate a combination of isoelectric focusing with immunodiffusion overlays.

The possibility of establishing very long concentration gradients may find uses not only in microbiology but also in various types of physicochemical studies.

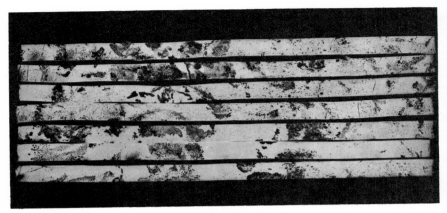

Fig. 19. "Bacteriogram" of open sandwich in Fig. 18.

With regard finally to clinical microbiology, the habits of the microbiologists and the time needed for computer programming may be the main factors limiting novel applications. It is likely that auxotyping of Enterobacteriaceae and gonococci will spread to other fields and that the agar-strip technique may find use also for *N. meningitidis* screening, which is obviously badly needed in such countries as Nigeria.

Acknowledgments

I am grateful to my colleagues A. Bolinder, H. Gyllenberg, and T. Illéni for many stimulating discussions; to S. Melkersson, L. Thelin, and L. Wegstedt for notable contribution to the design; to G. Goertz, L. Hellman, and B. Seger for invaluable assistance in the devlopment and manufacture of the equipment; and to I. Kühn and S. Lundell for performing the necessary microbiological and biochemical tests. However, the development would not have been possible without the financial aid of various granting agencies and the general support and stimulating interest of Biotec AB, which has given especially valuable advice with regard to the problems of patenting. In particular, the work was made possible through grants from the Marianne and Marcus Wallenberg Foundation, the Swedish Board for Technical Development, the Swedish Medical Research Council, and Nordforsk.

REFERENCES

1. E. A. Falch and C.-G. Hedén, *Ann. N. Y. Acad. Sci.*, **130**, 697, (1965).

2. C.-G. Hedén, Automation of Cultivation Techniques in the Rapid Identification of Microorganisms, Twelfth Pugwash Symposium on Rapid Detection and Identification of Microbiological Agents, Geneva, 1971.

3. C.-G. Hedén, Linear Cultivation, in Automation of Identification Procedures in Microbiology, *Bull. Ecol. Res. Commun.*, Stockholm, **17**, 174 (1973).

4. G. Vanzetti, F. Pucci, and G. Cosci, Automatic Analysis of Electrophoresis Strips by Means of a Cyclic Electronic Scanner, *Clin. Chim. Acta*, **20**, 215 (1968).

5. I. E. Buch, Automation of the Analyses of Urinary Steroids Using Quantative Chromatography and a Small Laboratory Digital Computer, *Clin. Chem.*, **14**, 491 (1968).

6. J. Winkelman and D. R. Wybenga, Automatic Calculation of Densitometer Scans of Electrophoresis Strips, *Clin. Chem.*, **15/8**, 708 (1969).

7. T. Illéni, G. Goertz, C.-G. Hedén, and L. Wegstedt, The Optical Measurement of Inhibition and Growth Zones in Bacterial Film on Agar Media. Abstracts. Symp. Rapid Methods and Automation in Microbiology Stockholm, June 3–9, 1973 (cf. preface).

8. G. C. Schild, M. Henry-Aymard, and H. G. Pereira, A Quantative, Single-Radial Diffusion Test for Immunological Studies with Influenza Virus, *J. Gen. Virol.*, **16**, 231 (1972).

9. C.-G. Hedén and G. C. Schild, unpublished results.

10. J. Kohn, A Preliminary Report of a New Gelatin Liquefaction Method, *J. Clin. Pathol.*, **6**, 249 (1953).

11. F. G. Martley, S. R. Jayashankar, and R. C. Lawrence, An Improved Agar Medium for the Detection of Proteolytic Organisms in Total Bacterial Counts, *J. Appl. Bacteriol.*, **33**, 363–370 (1970); O. S. Sandvik, Studies on Casein Precipitating Enzymes of Aerobic and Facultatively Anaerobic Bacteria, Thesis, Veterinary College, Oslo, Norway, pp. 42–48, 1962.

12. S. Arvidson, Hydrolysis of Casein by Three Extracellular Proteolytic Enzymes from *Staphylococcus aureus*, Strain V8, *Acta Pathol. Microbiol. Scand.*, Section B, in press.

13. N. Datta et al., R-Factors in *Escherichia coli* in Faeces after Oral Chemotherapy in General Practice, *Lancet*, **1**, 312 (1971)

14. B. Wretlind, G. Goertz, T. Illéni, S. Lundell, and B. Seger, Antibiotic Sensitivity Testing by Zone Scanning of Agar Strips, in C.-G. Hedén and T. Illéni, Eds., *Automation in Microbiology and Immunology*, Wiley, New York, 1974.

15. L. J. Hanka, Correlative Microbiological Assays, *Adv. Appl. Microbiol.*, **15**, 147–156 (1972).

16. A. E. Bolinder, Microbiological Plate Assay Methods for Free Amino Acid Levels in Blood, *Anal. Biochem.*, **27**, 370 (1969).

17. K. Carifo and B. Wesley-Catlin, Gonococcal Auxotyping: A System for Typing Based on Nutritional Requirements, *Abstracts*, M 334, Annual Meeting, American Society for Microbiology, 1973.

18 H. G. Gyllenberg and T. K. Niemelä, Basic Principles in Computer-Assisted Identification of Microorganisms, Chapter 13, this volume.

19. J. F. Stauffer, L. J. Schwartz, and C. W. Brady, Problems and Progress in Strain Selection Program with Cephalosporin-Producing Fungi, *Develop. Ind. Microbiol.*, **7**, 104 (1966).

20. A. Novick and T. Horiuchi, Hyperproduction of β-galactosidase by *Escherichia coli* Bacteria, *Cold Spring Harbor Symp. Quant. Biol.*, **26**, 239 (1961).

21. J. O. Lampen, N. P. Neumann, S. Gascon, and B. S. Montenecourt, Invertase Biosynthesis and the Yeast Cell Membrane, in *Organizational Biosynthesis*, H. J. Vogel, J. O. Lampen, E. V. Bryson (Eds.), Academic Press, New York, p. 363 (1967).

22. H. B. Woodruff, The Physiology of Antibiotic Production: The Role of the Producing Organism, *Symp. Soc. Gen. Microbiol.*, **16**, 22 (1966).

CHAPTER 3

Cultivating Microorganisms on a Substrate Tape

H. BÜRGER AND R. QUAST

SUMMARY

A continuous technique of cultivating and isolating microorganisms on a substrate tape has been developed. The cultivation tape consists of a semisolid medium on a transparent, flexible carrier tape. All types of agar media used in medical microbiological diagnostics can be applied without modification. Tapes are inoculated by a plating automate. About 180 fractional platings have been performed in one hour by an experimental prototype of the plating automate. Inoculated tapes may be rolled up and incubated. Besides fractional plating, plating for germ counts or antibiotics and for sensitivity tests can be executed on the tape. The tape may prove to be an indispensable part of future automatic microbiological diagnostic systems.

PROCEDURE

Microbiological diagnostics involves several laborious and often repeated manipulations (e.g., dilution plating of specimens to obtain single colonies, plating for germ counts, and plating for sensitivity tests). One can reduce the microbiological and clerical workloads of these three procedures considerably by mechanizing the plating of microorganisms and using a tapelike carrier for semisolid cultivation media.

Pilot studies demonstrate that agar layers can be coated approximately 2 mm thick onto a flexible carrier tape over a length of several meters. All types of semisolid agar media can be applied in the same way used in conventional petri dishes. The agar sticks to the perforated carrier tape so strongly that the tape may be rolled up and handled without the danger of separating the cultivation medium from the carrier. Rolls of coated cultivation tape can be stored easily in sealed plastic bags to prevent evaporation.

Dilution plating of bacterial suspensions was tested using an experimental prototype of the plating automate (1). In accordance with the usual microbiological routine, three loops were worked synchronously in the apparatus. The first loop took up a drop of the liquid specimen from a test tube and plated it in a meandering path on the surface of the agar layer, while the second loop passed through the lines made by the first and distributed the germs fetched en route in a similar way. The third loop repeated the steps of the second loop. Thus, depending on the bacterial density of the sample, single colonies were obtained after incubation along the lines of the second loop, or with very high bacterial densities along the lines drawn by the third loop.

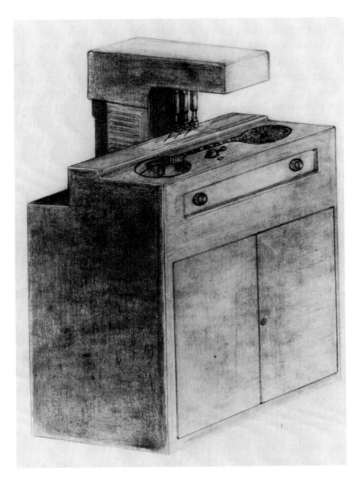

Fig. 2. The automatic plating machine.

APPLICATIONS

The first plating automate will be used for epidemic surveillance in a public health laboratory. For the purpose of screening for enteropathogenic bacteria—*Salmonella* and *Higella* sp.—Endo agar, Leifson agar, and if necessary Wilson-Blair agar will serve as semisolid nutrient media. The principal steps of the screening procedure are as follows. Specimens arrive in disposable tubes in which the free volume is limited by a stopper. Assigning

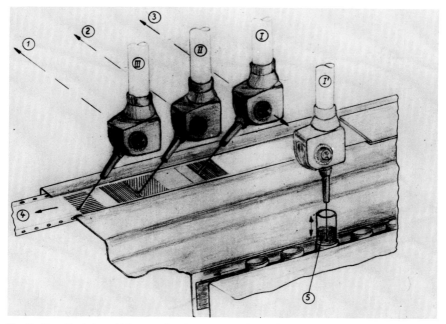

Fig. 3. Detail of the plating mechanism: I, II, and III, plating loops on rods; I′, loop I in sample takeup position; 1, 2, and 3, paths of the rods; 4, cultivation tape; 5, sample tube.

a specimen identification number and listing relevant patient data can be done in the usual manner or by an automatic data processing system. Before plating the specimen on the cultivation tape, enrichment of enteropathogenic bacteria is achieved in a conventional enrichment medium (e.g., selenite broth or tetrathionate broth). After incubation, the sediment in the tubes is resuspended by shaking, and the tubes are processed in the plating automate. The dilution platings on the cultivation tape produce single clones after incubation, and these clones can be scanned for suspicious colonies, either visually by a skilled technician or, in the future, by optoelectronic colony identification equipment. Such colonies can be easily isolated from the cultivation tape for subculture and for further biochemical and immunological identification. The isolation step may become automatic in the future, too, if an optoelectronic scanner, for example, controls the positioning of a loop mechanism similar to that just described.

ADVANTAGES

Among the benefits expected from the use of the nutrient tape in the immediate future are a reduction of approximately 85% in incubator space needed, a reduction of about 80% in weight of the material, and an approximate saving in nutrient medium of 60%, all compared with the results from conventional petri dishes. Additional advantages may be found in easier specimen identification, reproducibility, and quality of platings, and in a marked reduction of qualified manpower needed in the microbiological laboratory. It may be thought disadvantageous that the specimen to be plated must always be liquid and that sometimes swarming of, for example, *Proteus* bacteria may be encountered on some types of nutrient agar. However, specimens for dilution plating can be easily enriched in suitable media, as demonstrated in the screening procedure for enteropathogenic bacteria; moreover, specimens for germ counts usually are liquid themselves, and bacterial cultures for sensitivity tests also can be obtained in some liquid medium. The swarming of bacterial colonies is suppressed by incorporating p-nitrophenylglycerol into the semisolid medium. We have observed no other properties of the cultivation tape that might complicate the integration of this technique into the course of conventional microbiological procedures. Indeed, we expect the cultivation tape to become the main type of cultivation technique in future automatic colony detection, counting, and identification systems.

Acknowledgments

Our sincere thanks are due to Mr. K.-H. Wischniowski, who helped with the design and construction of the prototype apparatus and of the carrier tape, and to Mrs. A. Wischniowski and Mrs. B. Heitmann, who performed many of the pilot experiments. Also, we thank Prof. R. Siegert, director of the Hygiene Institute, for his kind interest and his support.

REFERENCES

1. H. Bürger and R. Quast, Neue Verfahren der Bakteriologischen Diagnostik, Eighth World Congress of Anatomic and Clinical Pathology, Munich, September 12–16, 1972.

CHAPTER 4

Parallel-Multiple Analysis: A Promising Tool for the Study of Complex Networks of Biochemical Functions as Represented by Living Cells

R. QUAST AND H. BÜRGER

SUMMARY

The system presented allows rapid measurements of the biochemical functions of living cells. The three essential parts—channel plate, pipetting device, and photometer—are described in detail. Also stressed is the importance of designing appropriate substrate combinations for the enzymatic function under test, with a reaction that detects the product resulting from the enzymatic action.

The system is primarily designed for collecting data on metabolic functions of living cells of procaryotes, usable for the calculation of a numerical taxonomy of procaryotes. Given such a taxonomy, diagnostic keys may be derived from a list of metabolic functions, which in turn may have applications in the chemistry and food industries.

INTRODUCTION

Procaryote cells may be characterized by metabolic functions only, because distinctive morphological features cannot be observed consistently in most of them. Although a vast amount of data on metabolic processes of procaryotes has been accumulated during the last decades, most studies pertain to possible biochemical reactions to be encountered in certain strains and to thermodynamic data on reaction mechanisms. These data also comprise metabolic pathways, whose sum may be regarded as a network of biochemical functions. In this chapter, rather than dwelling on further biochemical reactions that may occur in certain procaryote cells, eucaryote cells, or cell organelles, we wish to describe a systemic approach to the analysis of the networks of metabolic functions actually existing in living cells of genetically homogeneous populations. To do so, we must present some details of the apparatus to make clear the reasons for our certitude that the system to be described can meet the technical and methodological requirements inherent to the problem. The following demands must be fulfilled by an analytical system that is to be used in the investigation of such networks of biochemical functions in cells derived from a single clone:

1. The analytical system must be extremely versatile. To analyze the metabolism of a single type of cells, we need a great number of substrates and many reactions to detect the metabolic products. Machines built on principles of sequential analysis or having only a limited number of "channels" (i.e., programmed analytical tests) are not suitable. The system should be easily and speedily adaptable to new test reactions, giving the analytical chemist greatest freedom in choosing optimal test parameters.

2. The system must be computer compatible. The vast amount of data evolving from a moderately fast analytical system can be handled and evaluated only by electronic devices and by employing biostatistical techniques of data reduction.

3. The analytical reactions must be specific. Since the response of the detecting reaction must be unambiguously related to the metabolic reaction under test, it is clearly implied that chemically defined substrates must be used singly, and metabolic products must be identified by well-established chemical-analytical procedures.

4. The whole system must be realized within economical limits. By far the most important condition is economy of tests and apparatus, because the economical situation determines whether a system is employed.

DESCRIPTION OF PARALLEL—MULTIPLE ANALYSIS

Based on an invention of one of us (1), we designed a system of apparatus and chemical reactions (the latter being by no means complete) which fulfills the requirements of a versatile, specific, computer-compatible, and economic system for the rapid analysis of many single metabolic steps. The system is tentatively called PAVIAN from the German *Parallel-Vielfach-Analyse*, since it follows the principles of parallel-multiple analysis in which many samples are analyzed for certain substances, the chemical reactions taking place simultaneously.

The Channel Plate

Underlying the whole system is the channel plate (1), which comprises several trenches, each one divided by ridges. Thus reagent solutions can be delivered to the segments of the channel plate at one time, and tilting the plate will cause the solutions to flow together in each channel in a sequence determined by the position of the respective segment along its channel (Fig. 1). Finally, all reagents and the sample flow into a cuvette, and the mixture can be measured photometrically or fluorometrically. Channel plates of somewhat different design have been made of opaque glass; these were used in a series of pilot studies on single biochemical properties of bacteria (2–5) including mycoplasms (6), of yeasts (7), and leukocytes (not published). If necessary, the channels of the plate can be designed so that several multistep reactions can be performed in one channel to produce a single color reaction. The development of new detective reactions is facilitated by this property and by the fact that the sample together with the substrate may be placed and incubated in one of several segments of a channel. For microbiological work, disposable channel plates made of plastic appear to be most suitable.

Pipetting Devices

For optimal use of the channel plates, delivery of reagents into the segments must be done with the help of an automatic or semiautomatic pipetting mechanism. For this purpose, two different types of machine were designed, one of them having been employed in the pilot studies (2). It consists of a battery of overflow pipettes assembled according to the pattern of segments of the channel plates. In such an overflow pipette, a piston displaces the reagent fluid when immersed. The displaced volume drips directly from the

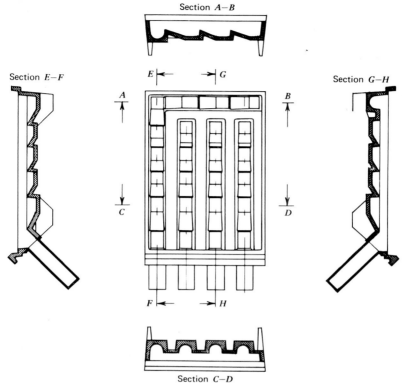

Fig. 1. Channel plate with adherent cuvettes. One L-shaped channel is accompanied by three straight channels of four segments each. The segments are designed to follow a regular pattern of 4×5 (crosses in center figure). Drawings of cross sections through the channel plate along the lines indicated appear around the center figure.

pipette tip into the segment of the channel below. Three positions of the piston provide for three fixed volumes that may be delivered. All pipettes of one battery are filled and emptied simultaneously. Thus, one channel plate is served at one stroke. The pipettes are filled simply by dipping them into a set of reagent tubes standing in predetermined order in a suitable rack and repeating the movement of the pistons in that position. Such a pipetting device is most economical, since it is purely mechanical and employs neither motor operated drive nor electronic controls. The volumetric precision is adequate only to semiquantitative work. This did not pose a serious problem in the pilot studies, however, since measurements of color intensity of the detective reactions could not be performed anyhow.

The second type of pipetting apparatus has been designed in rough outline. It will be more complicated, hence more expensive. On the other

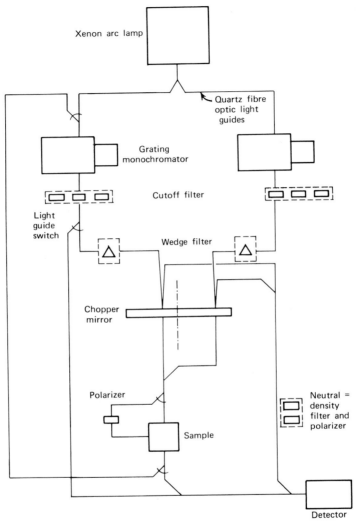

Fig. 3. Sketch of the optical part of the spectrophotofluorometer demonstrating the principal features of the instrument.

reflected and transmitted, each beam at its own frequency. Both transmitted beams as well as both reflected beams are assembled by uniting the respective light guides, and one of the light guides is led directly to the detector serving as energy reference for both monochromatic beams. The other is focused on the sample cuvette. A light switch in the course of the latter provides for absorption or fluorescence measurements, alternatively. Optical conditions for successful fluorescence determinations in turbid solutions are discussed by Udenfriend (8). Light from the cuvette is collected in a light guide, which leads to the detector. Thus the detector receives two pairs of monochromatic light beams in absorption measurements; alternately, either both energy reference beams or the monochromatic beams come from the cuvette. The electrical signals from the detector may be resolved for each of the wavelengths by electronic filters, because of the difference in chopping frequency. Electronic data handling equipment is used to convert the signals, store them, and calculate results. The whole spectrophotofluorometer is duplex controlled by a small computer. This enables the experimenter to employ absorption or fluorescence measurements at the appropriate wavelengths in rapid succession, as afforded by the program of reactions under study.

The design of the optical part provides for three different modes of operation: (1) absorbance measurements at two wavelengths simultaneously, (2) measurement of total fluorescence emission for two wavelengths of excitation simultaneously, and (3) measurement of fluorescence emission at one wavelength with one monochromatic excitation beam. All control and signal processing functions are planned for a rapid succession of samples (i.e., one sample measured each second). To achieve such a measuring rate, the disposable cuvettes fixed to the channel plates are continuously transported through the light beam by a conveyor belt mechanism. The transport mechanism is combined with a thermostated bath to regulate the temperature of the sample during measurement. The speed of the conveyor belt is controlled by the computer, too, and can be adjusted to the conditions of measurement. A cuvette can be stopped in the light path if spectra are to be taken in any of the three measurement modes or if moderately fast kinetics are to be registered. Slow kinetics are followed best by repetitive measurements conducted in preset time intervals.

CHEMISTRY

In our pilot studies more than 100 detective reactions were used (cf. Refs. 2–5), some of them especially designed for the conditions of parallel-multiple analysis. It is conceivable that the only reaction types employed were those in which a visible color marked the positive result, since we had to read the result by eye. That limitation of the scope of detective reactions will be overcome by the measuring instrument described. During the pilot experiments, we observed an extremely wide range of substrate conversion rates in microorganism. It is doubtful that detective reactions that comprise the whole dynamic range (i.e., 4–5 orders of magnitude) will ever be developed. Furthermore, colorimetry over a range of up to 4 or 5 absorbance units certainly does not follow Lambert-Beer's law. However, the concentration range of substrate converted can be roughly tested for in preliminary experiments and can be limited by choosing appropriate incubation times.

To fully exploit the potential of parallel-multiple analysis, many additional detective reactions must be designed. This task is facilitated by the number of different reagents that can be deposited in the segments of a channel, and by the versatility of the measuring instrument. Besides analytical-chemical color reactions, enzymatic reactions may serve in evaluating substrate conversion (9).

As we have pointed out, two types of reaction must be discerned. One kind is always the biochemical reaction under test; it occurs within the living cell during incubation. The other is the detective reaction, which is performed after incubation to detect the metabolic product formed. Biochemical reactions and detective reactions are mutually dependent, since only a single substrate or a substrate combination is added to the cell suspension.

Substrates can be metabolized over many subsequent steps or along different pathways. The first case can be analyzed by a set of detective reactions, each suitable to check for one intermediate product along the suspected metabolic chain. The other possibility can be tested for by using as substrates the metabolic products neighboring the presumed pool product in the chains and by analyzing for products following them. The strategy for analyzing metabolic chains in living cells rests mainly on similar procedures.

APPROACH TO A NUMERICAL TAXONOMY

A different strategy must be chosen for collecting data for establishing a numerical taxonomy of procaryotes. In contrast to the aforementioned procedures, this strategy can be compared with screening all possible metabolic functions by a fine-meshed multidimensional network of reaction pairs. Therefore, a great number of defined criteria must be tested for in each strain of cells. Calculations on the number of criteria have been performed by Sneath (10), who ended up with several hundred independent biochemical criteria for each strain. One can easily envisage that conventional techniques will never succeed in dealing with such an enormous number of tests, not even for strains of bacteria that are to be found in culture collections all over the world. It is also obvious that handling the vast amount of data can be done only with the help of computers and by means of some method of multivariate statistics presently developed and employed in psychology and the social sciences. It is improbable that methods of numerical taxonomy based on defined biochemical criteria will overturn our present taxonomial order of bacteria. On the contrary, the adopted order of microorganisms will be confirmed in most cases.

In a numerical taxonomy of procaryotes, we should expect the term "species" to be defined unambiguously; it should also be possible to determine for each strain the degree of divergence from an adopted species. Also, calculations will reveal whether the space between neighboring species is continuously filled in by intermediate strains or whether species are separated by well-defined boundaries. The question of whether all viable systems of combinations of metabolic functions are already realized by procaryote strains, can be tackled by employing the data collected by parallel-multiple analysis, together with the powerful techniques of calculation provided by electronic data processing systems.

PRACTICAL ASPECTS

Practical consequences can be derived only after accumulating enough data and finding answers to most of the scientific questions. Diagnostic keys for all kinds of applications will arise from the list of biochemical functions of cell strains when redundant data have been eliminated and when the criteria with the highest discrimination values have been defined. This will be the final step in the direction of an automatic biochemical identification of microorganisms. Besides that, the record of biochemical functions provides a catalog of great interest for the chemical industry, facilitating its search for inexpensive and specific catalysts for complicated synthetic processes.

Acknowledgments

Our sincere thanks are due to Mr. K.-H. Wischniowski for his help with the design and the construction of overflow pipettes and the new channel, and to Mrs. A. Wischniowski and Mrs. B. Heitmann for skilled technical assistance in the pilot studies.

Furthermore, we gratefully acknoweldge the unfailing interest and the helpful criticism of Prof. R. Siegert, director of the Hygiene Institute. The manufacture of the glass channel plates was made possible by a generous gift of Mr. Kurt Miether, Pforzheim.

This work was supported in part by the Deutsche Forschungs-gemeinschaft, grant Bu 263/1.

REFERENCES

1. German patent 1220083.

2. H. Bürger, *Zentralbl. Bakteriol. I. Abt. Orig.*, **196**, 469–476, (1965).

3. H. Bürger, *Zentralbl. Bakteriol. I. Abt. Orig.*, **202**, 97–109, (1967).

4. H. Bûrger, *Bakteriol. I. Abt. Orig.*, **202**, 395–401, (1967).

5. H. Bürger, W. Mannheim, and W. Stenzel, Ninth Conference on the Taxonomy of Bacteria, Brno, Czechoslovakia, September 24—27, 1969.

6. H. Bürger, M. Doss, W. Mannheim, and A. Schüler, *Z. Med. Mikrobiol. Immunol.*, **153**, 138–148 (1967).

7. H. Bürger, H.-L. Müller, and E. Pöppel, *Mycopathologia*, **34**, 241–252 (1968).

8. S. Udenfriend, *Fluorescence Assay in Biology and Medicine*, Vol. 2, Academic Press, New York and London, 1969, pp. 185–188.

9. H. Bürger and R. Quast, *Zentralbl. Bakteriol. Hyg. I. Abt. Orig. A*, **220**, 212–216, (1972).

10. P. H. A. Sneath, *Symp. Soc. Gen. Microbiol.* **12**, 290–332 (1962).

CHAPTER 5

Monitoring of Bacterial Activity by Impedance Measurements

AMIRAM UR AND D. F. J. BROWN

SUMMARY

Bacterial activity was detected by monitoring the changes in electrical impedance of broth cultures versus the impedance of sterile broth contained in an identical conductivity measuring cell. The signal, expressed as a curve, resembled growth curves. The impedance changes showed that the microorganisms metabolize substrate of low conductivity into products of higher conductivity.

Several species of bacteria (*E. coli, Klebsiella aerogenes, Pseudomonas aeruginosa,* staphylococci, and streptococci) and one of *Mycoplasma* (*My. argininii*) were studies. Characteristic curves obtained with different organisms in different media may allow rapid identification of pathogenic species.

The speed of response was proportional to the initial concentration of microorganisms. Signals were obtained within 2 hours with concentrations of bacteria as low as 10^3 to 10^4 per milliliter.

Preliminary experiments demonstrated the inhibitory effect of antibiotics on bacteria within 2 hours and indicated that the method may be useful for rapid determination of bacterial sensitivity to antimicrobial agents. It may also prove useful for the rapid assay of antibiotic levels in patients' serum. The method lends itself well to automation.

INTRODUCTION

Compared with other fields of clinical pathology, rapid and automated methods in microbiology have been slow to develop. One of the reasons for this is the relatively long reaction time of growing cultures. In clinical chemistry the procedure constitutes a substantial part of the time involved in a test, hence automation of techniques contributes considerably to the speed of completion of tests. In microbiology, where most tests involve incubation of growing cultures for at least several hours, automation of procedures would not have such merits and would serve only to save labor and increase the reliability of tests. Speeding up of microbiological tests would therefore depend on the development of more sensitive techniques for detection of the growth or activity of microorganisms. Apparently, however, further development of most of the techniques presently used would only marginally shorten the time involved. Thus new aspects associated with microbial activity need to be explored, one such aspect being the changes in electrical impedance which occur during the growth of bacterial cultures.

THEORETICAL BACKGROUND

The electrical resistivity of a medium containing active microorganisms changes with time in a way that indicates the possibility of using impedance measurements to monitor bacterial activity. A method employing this principle would have the immediate advantages of the high sensitivity offered by impedance measurements and simplicity of apparatus, as compared with such other methods of monitoring bacterial growth as photometric systems or the Coulter counter.

The impedance of an inoculated medium changes by about 4% from the time of inoculation to the point at which the stationary phase is reached. Such changes could be measured easily by using bridge circuits. Exploration of this field has been very limited, probably because of the difficulties associated with the detection of minute changes in impedance in an environment susceptible to noise and drift from evaporation of water, temperature variation, electrochemical reactions, and other processes that can mask the signal from active bacterial cultures. To facilitate such fine measurements, we had to overcome the noise and drift that are irrelevant to the activity of the microorganisms. Temperature variations, for example, affect the impedance by about 2% per degree centigrade. It is possible to reduce the effect of these processes to some extent by stabilizing the temperature and controlling evaporation, but such control is limited by practical considerations.

A different approach was taken in the present work (1). Instead of controlling these processes, their effect was canceled out by using a bridge circuit containing two measuring cells (Fig. 1) matched in physical dimensions. This system was originally developed for studies of blood coagulation (2,3). Both cells contain medium, but only one cell is inoculated with microorganisms; thus all impedance changes apart from those related to the activity of the microorganisms occur symmetrically, and undesirable signals, including those caused by broth contamination, cancel out. The signal relating to the activity of the microorganisms is therefore the only signal obtained, and it can be amplified to the desired level.

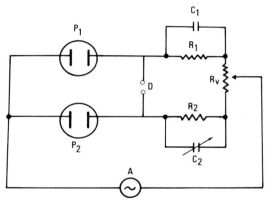

Fig. 1. The measuring bridge: P_1 = test cell; P_2 = reference cell; R_1 and R_2 = 15-k resistors; R_v = 10-turn 1-k potentiometer; C_1 = 5-pF capacitor; C_2 = 40-pF variable capacitor. An oscillator provides 1.5 V 10 kHz sinusoidal tension.

MATERIALS

The specially designed impedance measuring cells (Fig. 2) are made of glass capillaries with golden electrodes plated at each end. Since the cells are disposable, no cleaning and sterilization are required. The capillaries are stopped with plastic plugs to reduce evaporation. The samples are introduced by way of hypodermic needles inserted through the plastic plugs, whereupon the cells are placed in the cell holder of the measuring instrument (Fig. 3). The temperature in the cell holder is controlled to within 0.1 °C. The measured impedance changes are then continuously recorded on a chart recorder.

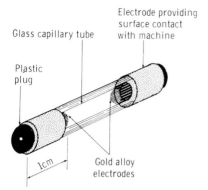

Electrode providing
surface contact
with machine

Glass capillary tube

Plastic
plug

1cm

Gold alloy
electrodes

Fig. 2. The comparative impedance monitoring cell (Stcatton and Co. Ltd., Hartfield, Hertshire, England).

Fig. 3. "Strattometer"—the comparative impedance measuring instrument.

RESULTS

A curve of the impedance changes of a culture of *E. coli* growing in PPLO broth appears in Fig. 4; its pattern is similar to that of the curve of viable microorganisms determined by the technique of Miles, Misra, and Irwin (4). Both curves are represented here on a logarithmic scale. The correlation of the two curves shows that the impedance changes follow the integral of the curve of viable organisms. When the culture reaches the stationary phase and the viable count is steady, the impedance curve continues to rise linearly (on nonlogarithmic scale), probably representing the continued metabolism of nonmultiplying or slowly multiplying organisms. The impedance reaches a constant level, presumably, when the culture dies.

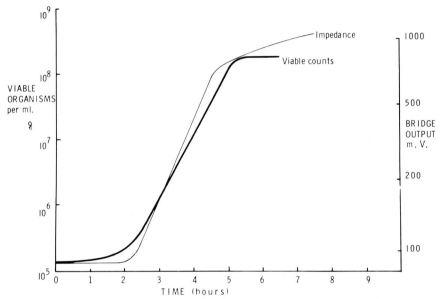

Fig. 4. Curves of changes in comparative impedance and the number of viable organisms in a culture of *E. coli* growing in PPLO broth.

This relationship indicates that the impedance changes are caused by an effect accumulative with time, such as the accumulation of metabolites from the microorganisms. In absolute terms, the impedance of the culture decreased continuously with time, indicating that the microorganisms metabolize substrates of low conductivity, such as carbohydrates, into products of higher conductivity, such as lactic acid. The effect of the impedance of the bacterial cells themselves is negligible at the low concentration of bacteria used in these studies.

The curve in Fig. 4 shows within 2 to 3 hours the activity of a culture in which the initial concentration was 10^5 microorganisms per milliliter. For practical, diagnostic, and experimental purposes, it is not necessary to record the entire curve, since the essential information is obtainable from the first part of the curve. Further developments have increased the sensitivity, allowing very rapid detection of bacterial activity. This was achieved by higher amplification of the signal and by the selection of media that accentuate the impedance changes caused by metabolism. Figure 5 contains curves of several organisms, the activity of all being detectable within 2 hours with inocula of 10^5 organisms per milliliter. As expected, *Pseudomonas aeruginosa*, an obligate aerobe, grew less well in this anaerobic system. The

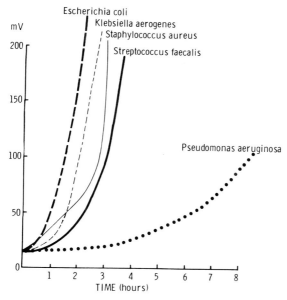

Fig. 5. Comparative impedance curves of different organisms grown in PPLO broth.

speed of response depends on the concentration of active microorganisms, as revealed in the impedance curves in Fig. 6. These curves show the response time to be proportional to the initial concentration of the microorganisms. A correlation curve could be constructed for any organism and used for rapid estimation of viable organisms in a culture. Figure 6 also shows the present sensitivity of the technique. At a concentration of 10^5 organisms per milliliter, the activity is detected within minutes. The activity of 10^3 organisms per milliliter is obvious within 2 hours and, since the measuring cell volume is only 0.1 ml, the method in fact detects the activity of only 100 microorganisms.

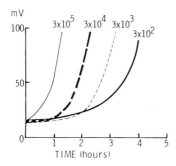

Fig. 6. The effect of inoculum growth curves of *E. coli* grown in PPLO broth.

OTHER APPLICATIONS

Another application of the method is in antibiotic-sensitivity testing. The results of such a test (see Fig. 7) were obvious within 2 hours; moreover, they are recorded automatically and quantitatively.

Because the method detects the metabolic activity of the microorganisms rather than their multiplication, it is possible to use it for detection of slow growing organisms or maybe even nonmultiplying organisms. One slowly growing microorganism successfully detected was *Mycoplasma argininii*. Its

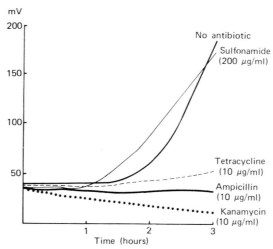

Fig. 7. The effect of antibiotics on impedance growth curves of a sensitive strain of *E. coli*. The false apparent resistance to sulfonamide is caused by sulfonamide antagonists in the medium.

activity was detected within 3 hours at concentrations of 10^5 microorganisms per milliliter. Extension of the method to the detection of slower-growing organisms, such as *Mycobacteria*, would depend on further improvement in the signal-to-noise ratio by selecting suitable media and controlling the noise.

The described advantages of rapid response time, high sensitivity, and automation were related here to tests on isolated microorganisms, such as in antibiotic-sensitivity tests. The same method is potentially applicable to the rapid identification of microorganisms. Different microorganisms in the same medium gave different impedance curves, and the same organism in different media gave different curves. It is possible, therefore, to prepare specific media for tests according to known biochemical reactions. This field of media engineering would require the combined "know-how" of microbiology, biochemistry, and electrochemistry to enable the preparation of media that would promote the growth of specific microorganisms while inhibiting the growth of others. At the same time, the media must provide substrates for specific reactions yielding optimal impedance changes. Suitable combinations of such tests, simultaneously performed, would allow the rapid identification of unknown microorganisms.

Acknowledgments

We thank Mr. David Norman and Mr. Nigel Lockyer for assistance with the instrumentation and Mr. Devendra Kothari for assistance with the bacteriological work.

REFERENCES

1. A. Ur and D. F. J. Brown, Detection of Bacterial Growth and Antibiotic Sensitivity by Monitoring Changes in Electrical Impedance, *I. R. C. S. Med. Sci.*, 1, 37 (August 1973).

2. A. Ur, Determination of Blood Coagulation Using Electrical Impedance Measurements, *Biomed. Eng.*, 5, 342 (1970).

3. A. Ur, Changes in the Electrical Impedance of Blood During Coagulation, *Nature*, 226, 269 (1970).

4. A. A. Miles, S. S. Misra, and J. D. Irwin, The Estimation of the Bactericidal Power of the Blood, *J. Hyg. Camb.*, 38, 732 (1938).

CHAPTER 6

Rapid Automated Bacterial Identification by Impedance Measurement

PAXTON CADY

INTRODUCTION

Microbiology has been affected relatively little by the rapid strides in technology occurring in other disciplines. The need for rapid detection and identification of microorganisms is obvious and urgent.

During the growth of microorganisms, as metabolism progresses, the composition of the supporting medium is altered as nutrients are converted by the organism into metabolic end products. Microbial catabolism of complex uncharged molecules, such as carbohydrates or lipids, results in their conversion into smaller charged units, such as lactic acid, acetic acid, or bicarbonate. Charged molecules such as polypeptides and proteins are converted by way of amino acids into ammonia and bicarbonate. As growth proceeds, all these processes tend to increase the conductivity of the supporting medium by the production of ion pairs from neutral molecules and by the reduction in size and resulting increase in mobility of large charged molecules. In a similar fashion, the dielectric constant increases as existing dipoles become smaller, new dipoles are formed, and inducible dipoles are created. Both these effects contribute to a decrease with time of the overall impedance of the medium as conductivity and capacitance increase.

Surprisingly, these effects have been little exploited by the microbiologist. However, conductimetric measurements of enzyme activity have been made by Lawrence and others (1), and slime organisms in paper mill effluents have been measured conductimetrically (2). Carbon dioxide in gases has been measured conductimetrically by Maffly (3); Ur refers to the applicability of impedance measures to microbiology in his studies of blood clotting (4); and recently Goldschmidt et al. presented a paper in which bacteria accumulated on a microfilter were detected impedametrically (5). The latter technique does not make use of the organism's metabolism, but rather detects the organism as charged particles after the filter is backwashed with conductivity water. This lack of application of impedance to microbiology may be partly due to the unavailability until recently of an inexpensive impedance bridge with precision greater than one part per thousand. However, the rapid advance of solid state electronics has removed this barrier.

In this chapter we propose that impedance measurements provide a rapid, inexpensive method of detecting microbial metabolism and that by the use of multiple, simultaneous measurements, rapid identification of microorganisms is possible within a few generations for a given microoogranism. Furthermore, we propose that by the use of specific inhibitors, certain

77

organisms can be detected in the presence of large numbers of commensals, and this may lead to the detection of specific organisms in a mixed flora without prior separation into pure cultures.

MATERIALS AND METHODS

Our measurement is made by conducting a signal variable in frequency (100–100,000 Hz) and in voltage (100–1000 mV) across two identical cells in series equipped with gold electrodes (see Fig. 1). Both cells are filled with the identical medium (liquid or semisolid), gassed with the appropriate gas phase when necessary, and stoppered. One is designated an experimental well and receives an aliquot of the bacteria to be tested. The other is designated the reference well and no additions are made to it. Both cells are placed in an incubator designed to hold its temperature variation to less than 0.1°C. Typically incubation temperatures are 37°C, but the incubator can be readily adjusted to higher or lower temperatures. A mechanical reciprocating shaker is used to assure adequate mixing of organisms and

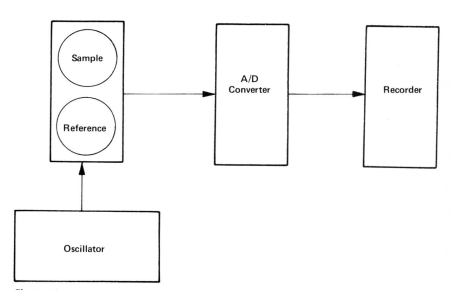

Fig. 1. Block circuit diagram for impedance measuring instrument (Bactometer).

nutrients and proper aeration. The impedance bridge actually measures the ratio of the impedance of the reference well (Z_r) to the impedance across both experimental and reference wells $(Z_r + Z_e)$. Initially this ratio is close to .500, since the impedance of one well is nearly identical to that of the other —that is, $Z_r/(Z_r + Z_e) = .500$. As the bacteria metabolize and the impedance in the experimental well decreases, the ratio increases. By the use of both a reference well and an experimental well, any effects on impedance occurring in both wells because of changes in temperatures, atmospheric pressure, stray radio signals, and so on, are canceled out.

The Bridge

The impedance bridge is made of an oscillator, which provides the signal at the selected frequency and selected voltage across both experimental and reference wells; a comparator and operational amplifier, which compares the impedance of the experimental well to the sum of the impedances of both experimental and reference wells; and a digitizing circuit, which by the use of an R-2 R resistance ladder gives a 16-bit binary number each time the impedance ratio of any cell pair is sampled. An address system sequentially samples the impedance ratio of any desired number of cell pairs, up to 64 pairs in the present instrument. Data can be displayed on a strip chart recorder by taking a portion of the 16-bit number to give a recording of impedance ratio against time. For low sensitivity only the initial portion of the digitized ratio is displayed. For highest sensitivity only the terminal portion of the digitized ratio is displayed. Alternatively, the entire 16-bit number can be processed by computer. We find that a chart speed of 15 cm/hour is convenient for most applications.

The Cell

The electrodes for the experimental and the reference cells are fabricated using conventional circuit board technology to provide a series of interdigitating gold-plated electrodes printed on glass epoxy circuit boards. These flat, printed electrodes form the floor or bottom of each experimental or reference chamber. The chamber is made of Lexan polycarbonate in the form of a cylinder; one end is glued to the circuit board, and the other end accepts a rubber stopper. The volume of each cell is approximately 2 ml. The rubber stopper provides a sealed top to prevent evaporation of liquid, contamina-

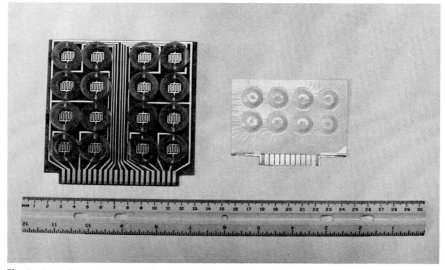

Fig. 2. Reusable and disposable modules (cluster of impedance measuring chambers).

tion, or loss of gas enrichments, and can be punctured by hypodermic needle to permit sterile additions and withdrawals to be made (see Fig. 2). A disposable cluster of chambers, which we refer to as a module, has been fabricated for us by the Rogers Corporation, Chandler, Arizona. This would be used in a fully automated system.

The Media

Conventional bacteriologic media are supplied by Difco, Detroit, Michigan, or BBL, Cockeysville, Maryland; these include trypticase soy broth (TSB), buffered azide glucose glycerol broth, Moeller KCN broth, urea broth, and Mueller-Hinton medium. In some instances special media were prepared. These include Thayer-Martin broth, which is identical to Thayer-Martin media in which the agar is omitted. For the detection of *Neisseria gonorrhoeae*, the medium of Morse was used (6). For the detection of *Streptococcus pyogenes*, a special medium was made that maintained red blood cells intact and also allowed the production of streptolysin O to occur. This was made by dissolving 25 g of veal infusion broth (Difco), 0.1 g of yeast extract (Difco), 60 g of sorbitol hydrate, and 0.075 g of cysteine hydrochloride in 1 liter distilled water, and adjusting the pH to 7.3.

The media in which the ionic content was minimized were made as follows: buffered azide glycerol broth consisted of 20 g of Neopeptone, 5 g of glucose, 0.5 g of sodium azide, and 5 ml of glycerol, brought to 1 liter with distilled water sterilizing at 116°C for 15 minutes.

KCN broth was prepared by dissolving 10 g of Neopeptone and 1 g of beef extract in 1 liter of distilled water, autoclaving for 15 minutes at 121°C and then adding 15 ml of 0.5% KCN. Mannitol broth was prepared by dissolving 1 g of beef extract, 10 g of Neopeptone, and 5 g of mannitol in 1 liter of distilled water and autoclaving for 15 minutes at 121°C. Lactose broth was prepared by dissolving 1 g of beef extract, 10 g of Neopeptone, and 5 g of lactose in 1 liter of distilled water and autoclaving 15 minutes at 121°C. Urea broth was made by dissolving 0.1 g of yeast extract, 0.091 g of KH_2PO_4, 0.095 g of Na_2HPO_4, and 20 g of urea in 1 liter of distilled water and sterilizing by filtration. Glucose broth was prepared by dissolving 7 g of Neopeptone and 5 g of glucose in 1 liter of distilled water and autoclaving for 15 minutes at 121°C. Gelatin broth consisted of 1 g of beef extract, 5 g of Neopeptone, and 120 g of gelatin dissolved in 1 liter of distilled water. Warming to 50°C facilitated the dissolving of the gelatin, but care was taken not to heat above 55°C, which might have charred the gelatin. Sterilization was for 15 minutes at 121°C.

When solid media were employed, it was important that the thickness of the layer of medium be 2 mm or less; otherwise diffusion of metabolites through the media to the electrodes was too slow to allow adequate detection.

Estimates of the number of bacteria in any given experiment were made by optical density measures in a Bausch&Lomb Spectronic-20 or in a Petroff-Hauser counting chamber when applicable, but always confirmed by appropriate dilution and plate counting. Plates in triplicate were inoculated with a known volume, usually 100 μl of the diluted sample. Five to eight sterile glass beads were then added to the plate, and rolling the beads back and forth produced a uniform dispersion of the inoculum. Less than 1% of the inoculum remained adherent to the beads. Plates from dilutions that yielded 30 to 300 colonies were counted.

Antisera

Antisera to *N. gonorrhoeae* absorbed against *N. meningitidis* and several other *Neisseria* species was kindly provided by Dr. William Peacock of the Center for Disease Control (7). Similar results were obtained by antiserum obtained from Sylvana Company in Millburn, New Jersey. Antisera to *E. coli* 0111:B4 were obtained from BBL.

Antibiotics

All antibiotics used were obtained from BBL in the form of Sensi-Discs—antibiotic-impregnated filter paper disks. To demonstrate the effect of antibiotics on bacterial growth, a single antibiotic-impregnated filter paper disk was placed in 1 ml of Mueller-Hinton broth. The antibiotic was thoroughly eluted from the paper disk, and aliquots of the antibiotic containing broth were used to fill experimental and reference impedance wells.

Bacteria

Many species were provided through the courtesy of Dr. Sidney Raffel and Mrs. Elizabeth Green of the Department of Medical Microbiology of Stanford University from subcultures originally obtained from American Type Culture' Collection (ATCC). Other named species were obtained directly from ATCC. Wild strains of *N. gonorrhoeae* were provided through the courtesy of Dr. Mary Riggs, Santa Clara County Venereal Disease Clinic, San Jose, California.

Impedance Versus Frequency Measures

Independent impedance measurements at varying frequencies were made by Dr. Leonard Nanis at Stanford Research Institute, Menlo Park, California.

Complement Fixation

All procedures were performed by the method of Levine and Van Vunakis (8). Freeze-dried complement was obtained from BBL; sheep erythrocytes and rabbit hemolytic antigen came from Hyland, Costa Mesa, California.

RESULTS

Single Measurements

The result of inoculating 10^8 *E. coli* into 0.5 ml of TSB (Difco) media appears in Fig. 3. Plate counts in triplicate were taken at intervals and compared with the impedance changes recorded by our instrument.

All the features of the classical bacterial growth curve can be identified in Fig. 3; the initial lag phase lasts about 40 minutes, followed by logarithmic

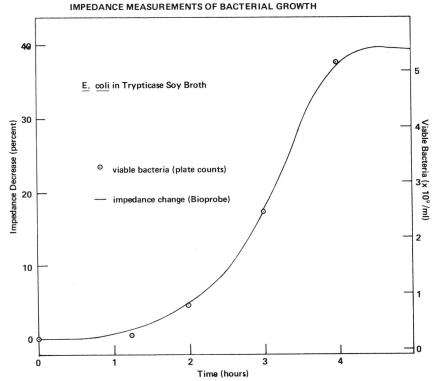

Fig. 3. Inoculation of 10^8 *E. coli* per milliliter in TSB. The solid curve represents impedance change (%) from initial impedance. The circled points represent the averages of 3 plate counts. Impedance was measured at 100 mV and 2 kHz.

growth to about 3 hours. Deaccelerating growth continues to about 4 hours, when the final stationary phase begins. The duration of each growth phase is determined by the number of organisms inoculated and the volume of "growing space" provided. In general, the smaller the volume and the larger the inoculum, the more rapidly stationary phase is reached. Thus one avenue to swift detection lies in the use of small volumes. The lag phase, by contrast, is more heavily dependent on the organism's past history.

The sensitivity of our method using commercially available media is indicated in Fig. 4. The logarithm of the number of organisms detected is plotted against the time of detection for different organisms.

The special case of *E. coli* grown in salt-poor medium is presented in Fig. 5. The use of salt-poor medium reduces the absolute conductivity of the

IMPEDANCE MEASUREMENTS OF E. COLI GROWTH AT VARIOUS FREQUENCIES

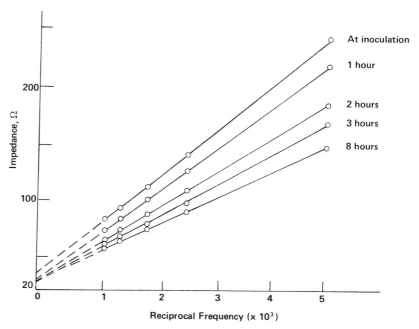

Fig. 6. Impedance measured at 100 mV at various frequencies. Impedance (Ω) plotted against reciprocal of frequency $\times 10^3$. Curves extrapolated to infinite frequency. Approximately 10^6 *E. coli* added to 0.5 ml of TSB.

The marked contribution of capacitance to the impedance change and the high values of the capacitance derived from our postulated model suggested that in addition to any changes in dielectric constant occurring in the medium, there may be deposition of a charged layer or film on the electrode forming a capacitor analogous to electrolytic capacitors. Evidence that such a layering does indeed take place is given in Fig. 7. *E. coli* is grown in TSB to stationary phase in an experimental well. The contents are carefully aspirated and the cell gently rinsed several times with fresh sterile medium; when impedance measurements again are taken with time, the impedance of the cell does not drop to its initial value but is nearly the same as the maximum value previously recorded. However, with time the value returns to initial levels. This process of returning the impedance of the cell to its initial value can be accelerated by mechanical scrubbing or washing or even the brief passage of alternating current.

We have only briefly investigated this capacitance-producing substance or substances but have observed that it is characteristic of *N. gonorrhoeae* as well as *E. coli*. In the case of *E. coli*, much of the substance is associated with the bacteria themselves, but some portion remains in solution when the bacteria are centrifuged. The substance is almost completely destroyed by freezing and thawing and by autoclaving.

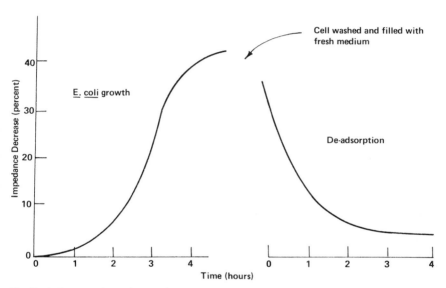

BACTERIAL METABOLITE DE-ADSORPTION FROM THE ELECTRODES

Fig. 7. Left curve: impedance change in percentage of initial value plotted against time in hours when 10^8 *E. coli* are grown in 0.5 ml of TSB. Impedance chamber then washed gently with sterile TSB, 0.5 ml of fresh TSB left in chamber and measurements resumed. Right curve: impedance change in percentage of original initial value with time in hours from resumption of measurement. Impedance measured at 100 mV and 2 kHz.

Multiple Measures

By measuring the impedance change associated with bacterial growth simultaneously in a number of differential growth media, characteristic patterns or fingerprints can be obtained for each organism. Figures 8 to 11 reveal the change in impedance with time in seven differential media, for four different organisms. Note the U-shaped dip in the impedance in most of the curves presented, which occurs during the first 10 to 20 minutes. This we feel is due to thermal effects as the module (cluster of impedance wells) warms up to oven temperature from room temperature. Prewarming the module served to reduce this effect but never completely eliminated it.

There are several discrepancies between metabolisms observed impedemetrically and conventionally. In Fig. 8, for example, *E. coli* shows growth in KCN broth where it normally fails to thrive. However, the impedance change for *E. coli* occurs later and is of lower magnitude than that of *Klebsiella* in KCN broth. Again, in Fig. 9, *Klebsiella* is seen to metabolize gelatin. Here the discrepancy may be due to the elevated temperature. Tests of gelatin hydrolysis must be conducted at 22°C; otherwise the gelatin melts and any liquefaction of gelatin is hard to see. We performed all our tests at 37°C; hence it is possible that *Klebsiella* utilizes gelatin at this elevated temperature even though it does not at 22°C. In Fig. 10 *Proteus* is seen to utilize mannitol, and in Fig. 11 *Enterobacter* is utilizing lactose medium. In both cases, though, the media contained sufficient nutrients other than the designated sugar to sustain metabolism for several generations. In later experiments, lactose medium devoid of any peptone and containing only lactose and a trace of beef extract gave more reliable correlations with conventional established metabolic behavior.

We conclude that characteristic and useful short-term metabolic patterns can be established by impedance measures even if these patterns are not exact replicas of conventional 18 to 24 hour metabolic patterns. We propose that these or analogous metabolic behavior patterns may serve as a rapid identification procedure for bacteria commonly found as a single pathogen.

Response to Antibiotics

We have tested the effect of approximately one dozen antibiotics on approximately one dozen species of microorganism. A typical result appears in Fig. 12, where approximately 10^6 organisms per milliliter of *Salmonella enteritidis* were inoculated into Mueller-Hinton broth with no antibiotics and into Mueller-Hinton broths containing 20 μg/ml of penicillin, 60 μg/ml of chloramphenicol, 60 μg/ml of tetracycline, and 20 μg/ml of gentimycin, respectively. The impedance change with time is shown for each antibiotic and for the antibiotic-free control. The slopes of each curve are well established and nearly constant after 30 minutes. In each case there is clearly a reduced rate of impedance change and, by inference, a reduction in metabolism in the presence of the antibiotic. This specific strain of *Salmonella enteritidis* was demonstrated by the Kirby-Bauer technique to be penicillin intermediate and sensitive to chloramphenicol, tetracycline, and gentimycin.

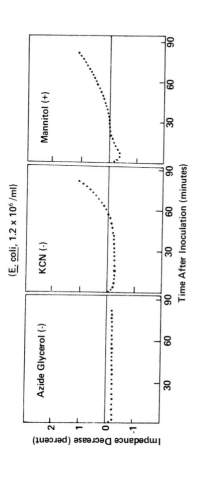

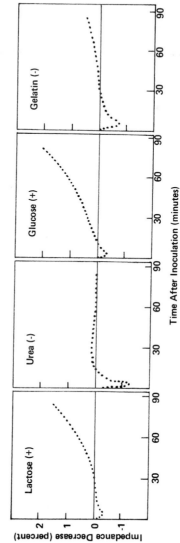

Fig. 8. Identification of *E. coli* by characteristic behavior in various media. Media are described under "Materials and Methods." Impedance change is in percentage of initial value; time is in minutes; The symbol (+) indicates that the organism shows visible positive growth at 18 hours in the medium. The symbol (−) indicates normally no visible growth at 18 hours. Impedance is measured at 100 mV and 2 kHz. Volume of each medium is 0.5 ml.

MICROORGANISM IDENTIFICATION BY RECOGNIZING CHARACTERISTIC BEHAVIOR IN VARIOUS MEDIA

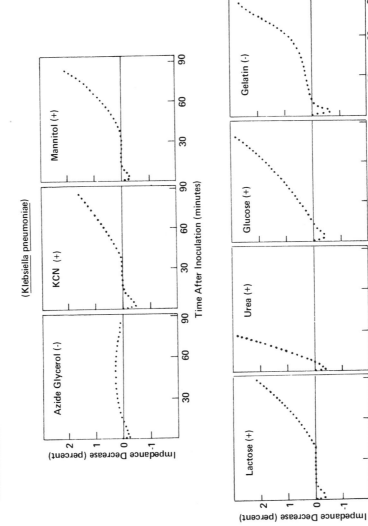

Fig. 9. Impedance changes measured as in Fig. 8 for *Klebsiella pneumoniae*.

MICROORGANISM IDENTIFICATION BY RECOGNIZING CHARACTERISTIC BEHAVIOR IN VARIOUS MEDIA

(Proteus mirabilis, 10⁷/ml)

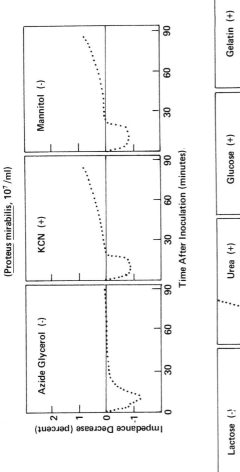

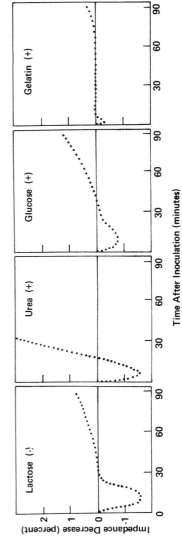

Fig. 10. Impedance changes measured as in Fig. 8 for *Proteus mirabilis.*

91

MICROORGANISM IDENTIFICATION BY RECOGNIZING CHARACTERISTIC BEHAVIOR IN VARIOUS MEDIA

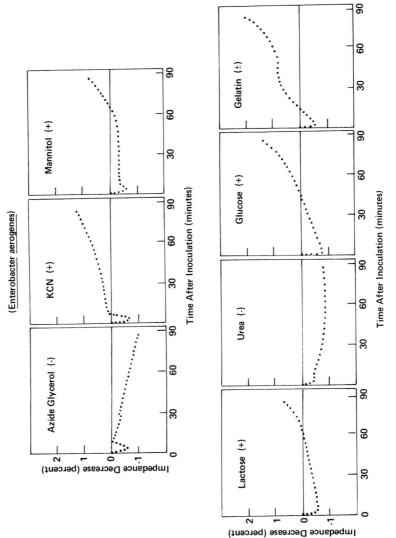

Fig. 11. Impedance changes measured as in Fig. 8 for *Enterobacter aerogenes*.

92

BACTERIAL RESPONSE TO ANTIBIOTICS

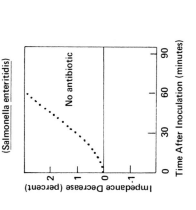

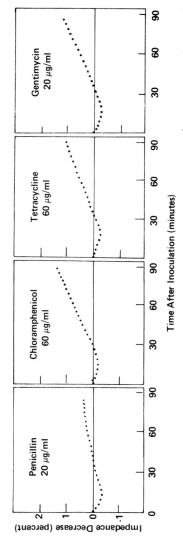

Fig. 12. Effect of antibiotics on bacterial growth. *Salmonella enteritidis*, approximately 10^6/ml, was placed in 0.5 ml of Mueller-Hinton broth. Impedance change is in percentage of initial value; time is in minutes. Antibiotics obtained from antibiotic-impregnated filter paper disks. A single disk of each antibiotic shown was eluted into 1.0 ml of Mueller-Hinton broth, which was then used to fill reference and experimental wells, 0.5 ml each. Impedance was measured at 100 mV and 2 kHz.

CHAPTER 7

Bacterial Identification by Microcalorimetry

W. J. RUSSELL, J. F. ZETTLER, G. C. BLANCHARD, E. A BOLING

SUMMARY

Microcalorimetry provides a means for revealing obscure metabolic phenomena of growing bacterial cultures. Our preliminary report (6) described the differentiation among representative Enterobacteriaceae through their elaboration of uniquely descriptive and reproducible heat profiles.

The capability of this approach to identification has been further characterized through examination of more than 200 cultures representing 47 species or types from 24 genera of clinically significant microorganisms. Virtually all are identifiable by microcalorimetry in our present system with a single medium.

Most organisms typically produce maximum outputs of 40–60 μcal/(sec) (ml) and return to baseline in 5 to 7 hours thereafter. Some profiles are identifiable within 3 hours of onset of heat production, whereas others require as long as 14 hours. Strain differences are detectable by this method, and the relationship of known biochemical differences with profile alterations has been initially investigated.

The microcalorimetry instrument has been designed for clinical microbiology application. It simultaneously accommodates up to 50 samples and utilizes an integral computer system for data acquisition and processing, as well as for instrument control.

INTRODUCTION

The continuous thermal monitoring of biological systems provides a means for detection of subtle changes in metabolism. Microcalorimetric analysis has been especially useful in the areas of ecology (1) and comparative physiology, where alterations in thermogenesis may be related to senescence, metamorphic transformation, and environmental variation (2).

The specific application of calorimetry to the study of microorganisms is described by a number of early workers. Hill, in 1911, investigated the action of yeast cells on cane sugar by microcalorimetry (3). Later, Bayne-Jones and Rhees reported on the relation of heat production to phases of bacterial growth (4). With the advent of more satisfactory techniques and instrumentation in recent years, the studies of Forrest (5) and Prat (2) are particularly noteworthy. Forrest's work dealt with the effects of growth-limiting energy and nutritional factors on thermogenesis, whereas Prat demonstrated that certain bacteria generate a fluctuating but specific and reproducible heat profile when rate of heat production is plotted against time.

In 1969 Instrumentation Laboratory, Inc., began seeking new approaches to the automation of microbiology, with special emphasis on clinical microbiology applications in which many specimens are submitted daily for bacterial identification. Microcalorimetric analysis appeared to offer distinct advantages in this area. Thus several prototype microcalorimeters were designed and tested specifically to explore whether bacterial heat profile reproducibility and profile fine structure could be used to differentiate one microbial species from another. Our preliminary report in *Nature* described characteristic heat profiles for different members of the family Enterobacteriaceae (6). The obvious implication was that organisms can potentially be identified in a relatively short time and with little effort.

This report describes our continuing exploration of bacterial identification by microcalorimetry and also improved instrumentation for clinical microbiology application. The scope of our work has been confined to microbial cultures of clinical significance, with major emphasis on the Enterobacteriaceae, *Pseudomonas*, and other gram-negative rods. Gram-positive cocci, such as *Staplylococcus* and *Streptococcus*, yeasts and others, have been studied to a lesser extent.

EXPERIMENTAL

The current Instrumentation Laboratory microcalorimeter is an isothermally jacketed, 50-channel device operating at 37°C (Fig. 1). It is an

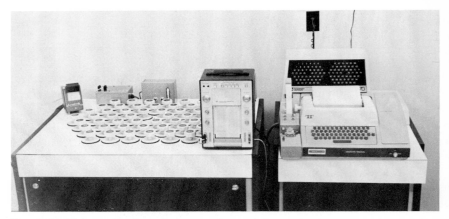

Fig. 1. Multichannel microcalorimeter: lower left quadrant, constant-temperature chamber; left center, series of 50 numbered knobs, each constituting a cover for an individual channel; upper right quadrant, indicator lights showing channels in use; lower right quadrant, computer; right center, printout and keyboard for directing computer functions; center, recorder for displaying thermal data in analog form.

engineering model and not yet commercially available. An integral computer system is used for data acquisition, processing, and display, as well as for complete instrument control and monitoring. The long-term baseline stability is 0.8 μcal/(sec)(ml) over a 24-hour period. This value amounts to 2% of maximum outputs of approximately 40 μcal/(sec)(ml) produced by most organisms.

The pure culture is grown undisturbed in the calorimeter in a short screw-cap tube charged with 4 ml of liquid media (Fig. 2). After testing of several commonly used enrichment media, brain-heart infusion broth from a single manufacturer's lot was selected. The inocula consisted of approximately 500 cells obtained by dilution of a logarithmically growing broth culture.

The device is convenient to use. The inoculated sample is manually introduced into a preselected individual calorimeter channel, and the outer cover is replaced. Then, by a simple computer command given through the console, the instrument assumes control, bringing the sample to baseline temperature and continuously monitoring and storing the temperature data with no further operator attention. It should be noted here that any or all of the 50 channels can be operated at any given time, and experiments can be initiated or terminated in any of the channels without disturbing ongoing experiments in adjacent channels. Additionally, a printout report can be obtained at any time on any or all channels, thus allowing the operator to

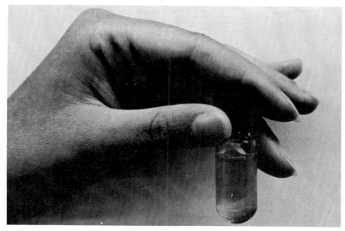

Fig. 2. Sample vessel for calorimetric analysis with 4 ml of growth media.

escape the highly confusing task of following, mentally or by logbook, the status of the analysis in each channel.

Since the organism heat profile is the parameter used for identification, the curve information stored in the computer for each channel is plotted on a strip chart recorder through a digital-to-analog interface on appropriate command. At present, visual comparison of the experimental heat profiles with a catalog of known profiles is used for differentiating among the organisms. Ultimately, it is envisioned that the computer will contain a library of standard profiles and a profile-matching program that is rapid and eliminates subjective comparisons by the operator.

The heat profile generated by *Enterobacter cloacae* is depicted in Fig. 3 to illustrate profile complexity. For this organism, a maximum output of approximately 35 μcal/(sec)(ml) is reached about 3 hours after onset of heat production, after which there are two major changes in heat generation before the return to baseline, approximately 7 hours later. With this organism, as well as with other Enterobacteriaceae tested, between 10^5 and 10^6 organisms per milliliter are needed to initiate detectable heat production in the present device. The same plotting scales are used for all other heat profiles given in this report.

RESULTS AND DISCUSSION

To demonstrate both the biological reproducibility of the heat profile and the high degree of uniformity among the individual microcalorimeters in the

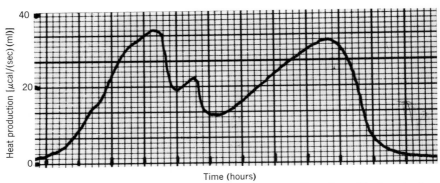

Fig. 3. Representative heat profile (*Enterobacter cloacae*). In Figs. 3 to 10, the abscissa represents time, and the ordinate represents heat production.

instrument, Fig. 4a depicts profiles generated by the same isolate of *E. coli* grown simultaneously in 12 different channels of the microcalorimeter. These heat profiles are essentially superimposable. In Fig. 4b four of the profiles from the Fig. 4a are plotted on an expanded scale to allow more adequate visualization of the reproducibility. The profiles are essentially identical in all respects. On close examination, however, slight variations are detectable at the areas of peak heat production and midway on the descending side of the profiles. In a way, these minor variations are reassuring features, since we are dealing with complex biological phenomena and would expect to obtain some variability. The timing and amplitude of each

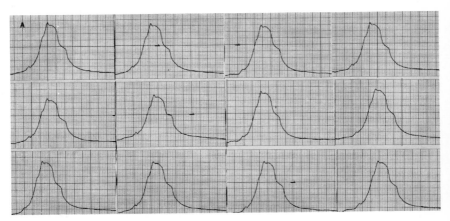

Fig. 4. Heat profile reproducibility: (*a*) same isolate of *E. coli* grown simultaneously in 12 separate channels; (*b*) four representative profiles from (*a*) plotted on an expanded ordinate scale.

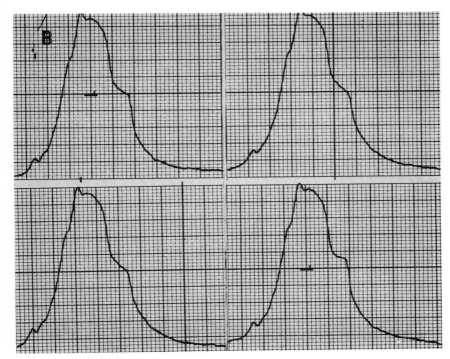

Fig. 4. (Cont.)

varying thermal event are virtually identical. These profiles, as well as all others presented, are photographs of the actual plots rather than artists' redrawings.

Additionally, an identical follow-up experiment was conducted on this organism after storage on trypticase soy agar at 4°C for 50 days. The profiles were the same as before. We do, however, recognize the possibility of mutational or adaptive changes occurring in older cultures, and we are lyophilizing the cultures that will be used as so-called standard profile organisms.

To illustrate the magnitude of profile diversity being considered, Fig. 5 shows heat profiles among representatives of the family Enterobacteriaceae. Figure 6 contains profiles of representative organisms of clinical significance other than the Enterobacteriaceae. These cultures were obtained either from the U.S. Public Health Service Center for Disease Control, as reference cultures from diagnostic laboratories or as fresh clinical isolates. The identity of each culture was verified by conventional biochemical and/or serologic tests before calorimetric analysis.

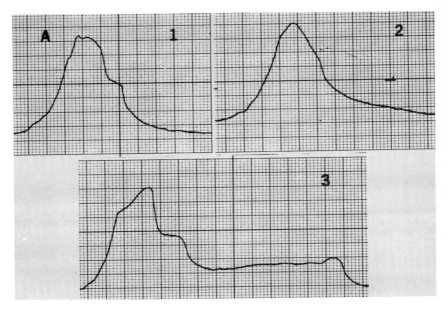

Fig. 5. Heat profiles among representatives of the family *Enterobacteriaceae*: (*a*) (1) *Escherichia coli*, (2) *Shigella sonnei*, (3) *Edwardsiella tarda*; (*b*) (1) *Citrobacter*, (2) *Enterobacter cloacae*, (3) *Enterobacter hafniae*, (4) *Providencia stuartii*; (*c*) (1) *Proteus morganii*, (2) *Proteus mirabilis*, (3) *Proteus rettgeri*, (4) *Proteus vulgaris*; (*d*) *Salmonella*: (1) group I, (2) group D, (3) group E, (4) group C; (*e*) (1) *Serratia marcescens*, (2) *Klebsiella pneumoniae*.

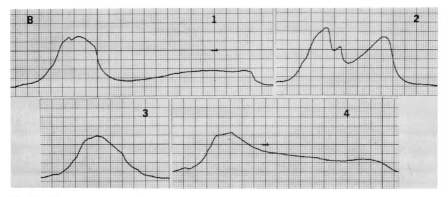

Fig. 5. (Cont.)

110

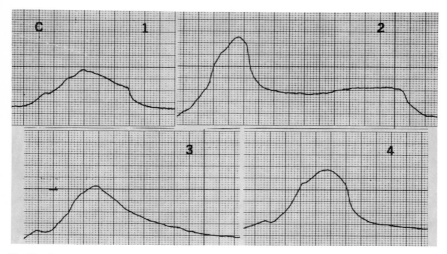

Fig. 5. (Cont.)

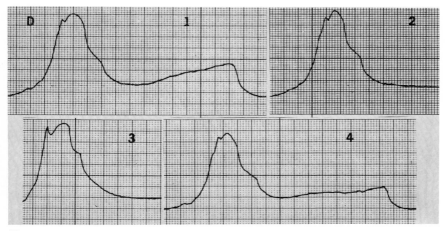

Fig. 5. (Cont.)

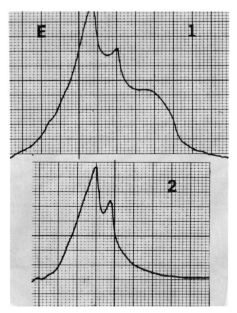

Fig. 5. (Cont.)

Thus far, we have examined more than 200 microbial cultures of clinical significance, representing 24 genera and 47 species or types. Each organism was found to yield reproducible heat profiles. Although some profile shapes are basically similar among certain groups, the amplitude has invariably been different; thus virtually all organisms have been distinguishable by calorimetry. There are, however, two exceptions, the most perplexing being the generation of nearly identical profiles by two distinctly different organisms—*Escherichia coli* and *Salmonella paratyphi B*. However, these two organisms showed entirely different heat profiles when grown in another medium (dextrose broth, Fig. 7). Second, the *Salmonella* examined (only 9 cultures so far, from five serologic groups) have not demonstrated any identifiable thermal patterns among the serologic groups.

In terms of clinical microbiology application, our microcalorimetric approach offers a promising rapid and especially simple method for microbial identification. Some organisms generate profiles with unique characteristics on the ascending side of the major peak, enabling identification within 3 hours of onset of heat production. Most organisms, however, require that the profile be generated for 7 to 10 hours before identification can be made. In all cases observed thus far, the heat production ceases, as seen by a return to baseline before 24 hours has elapsed.

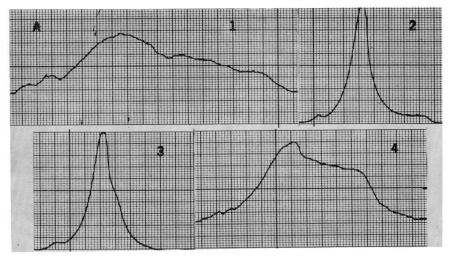

Fig. 6. Heat profiles of other clinically significant microorganisms: (*a*) (1) *Staphylococcus epidermidis*, (2) α-*Streptococcus*, (3) *Streptococcus faecalis*, (4) *Staphylococcus aureus*; (*b*) (1) *Pseudomonas maltophilia*, (2) *Bordetella bronchiseptica*, (3) *Pasteurella multocida*, (4) *Pseudomonas aeruginosa*; (*c*) (1) *Listeria monocytogenes*, (2) *Candida tropicalis*, (3) *Alcaligenes odorans*, (4) *Aeromonas hydrophila*.

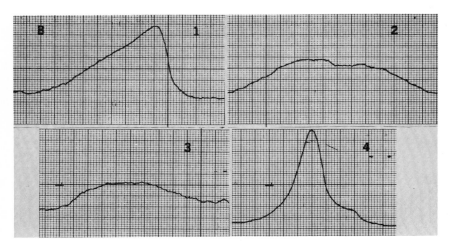

Fig. 6. (Cont.)

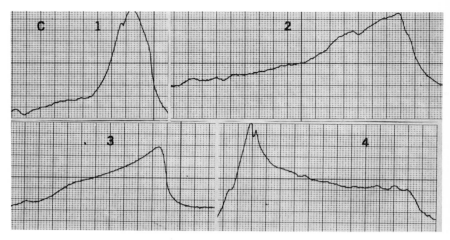

Fig. 6. (Cont.)

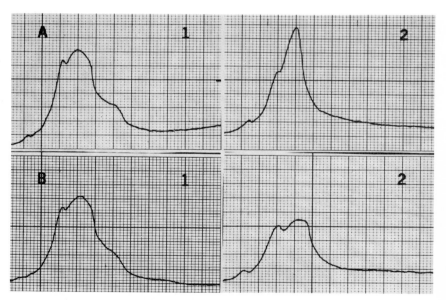

Fig. 7. Comparison of heat profiles generated by *Escherichia coli* and *Salmonella paratyphi* B in two different culture media. (*a*) *Escherichia coli*: (1) brain-heart infusion broth, (2) dextrose broth; (*b*) *Salmonella paratyphi* B: (1) brain-heart infusion broth, (2) dextrose broth.

114

Obviously, these heat profiles represent a thermal composite of a number of sequential and simultaneous metabolic events. The studies by Forrest demonstrated that the ascending side of the thermogram was closely associated with exponential growth of the organism and that heat production ceased abruptly following the exhaustion of the energy source by certain organisms (5). However, virtually nothing is known of the events causing the relatively small and sudden changes in heat production at many successive points along the profiles.

Although it is now possible to differentiate among the organisms examined thus far, it became apparent that different isolates of the same species sometimes generate different heat profiles. This was especially true among *E. coli*. From a practical standpoint, we wanted to establish experimental guidelines aimed at preventing the library of standard profiles from becoming excessively large without sacrifice of reliability. Recently we began testing the hypothesis that there is a relationship of *known* physiological differences within a particular species with basic heat profile shapes. Biochemical tests that are routinely employed in the clinical microbiology laboratory for identification were used to test this hypothesis.

This relationship is exemplified in the case of six different cultures of *Enterobacter hafniae*. Two basic profile shapes have been observed—one showing a sudden cessation of heat production (Fig. 8*a*) and the other exhibiting a more gradual decrease in heat production (Fig. 8*b*). These organisms are the same biochemically except for one major feature. Those in Fig. 8*a* are Voges-Proskauer positive and those in Fig. 8*b* are Voges-Proskauer negative.

A somewhat different situation occurs with *E. coli*, the organism demonstrating by far the greatest profile diversity within a species. Figure 9 shows heat profiles of 13 different *Escherichia coli* cultures for which there are known biochemical differences among ornithine and lysine decarboxylase and arginine dihydrolase tests. All cultures were lactose and indole positive and hydrogen sulfide, Voges-Proskauer, and citrate negative. The four cultures in Fig. 9*a* have the same biochemical test results, and all have the same basic profile shape, at least for the first major heat-producing event. The two profiles in Fig. 9*b*, although differing in their arginine reaction, demonstrate the same profile shape and are similar to those in Fig. 9*a*. At this point, additional profile shapes emerged in Fig. 9*c*; four completely different shapes are observed among organisms having identical but biochemical test results in different combinations from those shown previously. The three profiles in Fig. 9*d* were generated by cultures with still a different combination of test results; again, they demonstrate no uniformity of shape with known biochemical variation. The center profile, however, does resemble those of Figs. 9*a* and 9*b*. The conclusion is obvious in the case of these different *E. coli* cultures—although a predominating profile shape

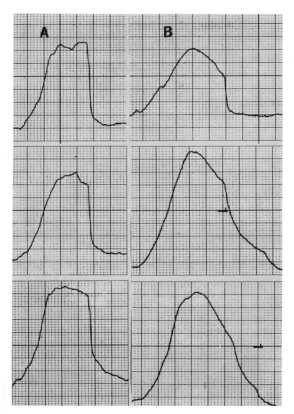

Fig. 8. Profile diversity among different *Enterobacter hafniae* cultures: (*a*) Voges-Proskauer (+), (*b*) Voges-Proskauer (−).

exists, there is no apparent relation of profile shape to known biochemical variation. It is recognized that these biochemical characterizations are reflecting only small evidences of the myriad physiological transformations that produce the thermal changes. It would be a monumental challenge to relate each thermal event to biochemical reactions.

Profile diversity within a species has not, however, been a prevalent feature in the present study. It should be emphasized that most organisms demonstrate remarkably uniform responses among different isolates of the same specie. *Klebsiella pneumoniae* furnishes a typical example. Heat profiles generated by six different isolates of this organism appear in Fig. 10. No known biochemical variation was observed, and all have the same basic profile configuration.

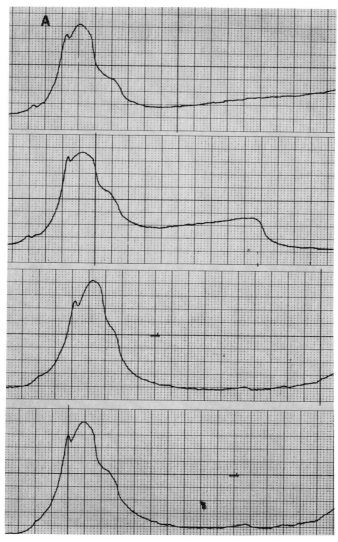

Fig. 9. Profile diversity among different *Escherichia coli* cultures: (*a*) ornithine (+), lysine (+), arginine (−); (*b*) top, ornithine (−), lysine neutral, arginine (−); bottom, also ornithine (−), lysine neutral, but arginine (+); (*c*) ornithine (+), lysine neutral, arginine (+); (*d*) ornithine (+), lysine neutral, arginine (−).

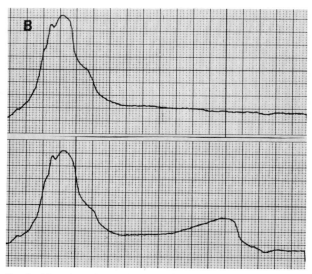

Fig. 9. (Cont.)

A number of parameters of investigation are planned for the future. Foremost will be the examination of many more strains of organisms to increase the reliability of identification. Another area is the development of culture media to optimize differential heat profiles, with special emphasis on organisms that do not grow well in broth cultures because of special metabolic needs. Also included is the exploration of unique methods for equating physiologic functions with both thermal uniformity and diversity. Additionally, heat profile generation may be useful in probing taxonomic problems by characterizing unique features of a given population of microorganisms.

Acknowledgments

The authors wish to acknowledge their indebtedness to Ms. Suzanne R. Farling for her diagnostic microbiology expertise and as instrument operator, and to Mr. Andrew Capitula for his invaluable contributions to instrumentation design and construction.

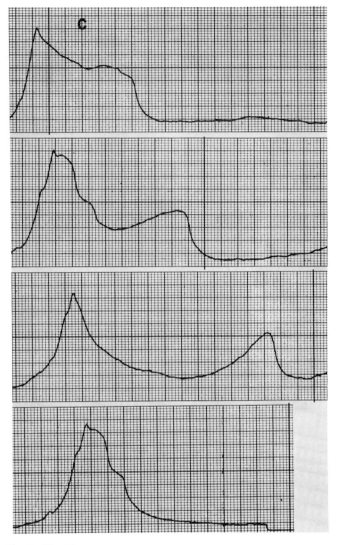

Fig. 9. (Cont.)

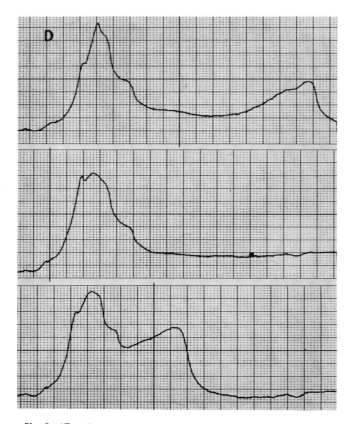

Fig. 9. (Cont.)

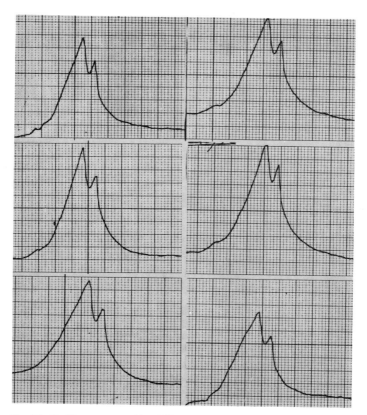

Fig. 10. Profiles generated by different cultures of *Klebsiella pneumoniae*.

REFERENCES

1. U. Mortensen, B. Noren, and I. Wadso, *Bull. Ecol. Res. Commun.* (*Stockholm*), **17**, 189 (1973).
2. H. Prat, in *Recent Progress in Microcalorimetry*, Macmillan, New York, 1963, p. 111.
3. A. V. Hill, *J. Physiol.*, **43**, 261 (1911).
4. S. Bayne-Jones and H. S. Rhees, *J. Bacteriol.*, **17**, 123 (1929).
5. W. W. Forrest, in *Biochemical Microcalorimetry*, Academic Press, New York, 1969, p. 165.
6. E. A. Boling, G. C. Blanchard, and W. J. Russell, *Nature*, **241**, 472 (1973).

CHAPTER 8

Rapid Automated Identification of Microorganisms in Clinical Specimens by Gas Chromatography

BRIJ M. MITRUKA

SUMMARY

Gas chromatographic (GC) techniques were employed for identification of microorganisms in clinical specimens in the following ways: (1) detection of one or more characteristic products of microorganisms *in vitro* and *in vivo*, (2) comparison of metabolic profiles (fingerprints) of organisms in infected and uninfected specimens, and (3) analysis of the products of microbial isolates after a brief incubation in an enriched broth culture medium. The procedures of fractionation, solvent extraction, derivative preparation, and sample injection were automated for rapid analyses. Chromatographic data obtained by three columns and four detectors, operating simultaneously, were analyzed by an on-line integrator and tape recorder system. The chromatographic patterns were compared either manually or using a digital computer.

Gas chromatograms of serum samples from rats infected with a single strain from the genera *Salmonella, Pseudomonas, Escherichia, Proteus, Serratia, Haemophilus, Streptococcus, Staphylococcus,* or *Diplococcus* yielded 2 to 8 characteristic peaks that were readily discernible. The compounds associated with the *in vivo* metabolic activity of the individual bacterial species were identified as organic acids, primary amines, sugars, and glycoprotein components. Characteristic GC profiles of human infections were obtained for bacterial infections due to mycobacteria, pneumococci, staphylococci, or certain gram-negative bacilli. By analysis of chromatograms of 300 coded sera from infected patients, the infective agents were rapidly identified with 97% accuracy. Gas chromatographic data of organic acid products from 200 different bacterial strains were digitalized and stored in a computer for rapid analysis. The identification of unknown bacterial isolates within these standards was achieved in approximately 4 hours. The GC analysis of sera or tissue samples from mice infected with vesicular stomatisis virus (VSV), influenza virus, or herpesvirus showed a characteristic metabolic profile for each type of infectious agent tested. These studies suggest that gas chromatography is a potentially useful tool for the rapid automated identification of microorganisms in clinical specimens.

INTRODUCTION

The identification of microorganisms in natural environment or in clinical specimens requires isolation of the infective agents followed by a battery of morphological, serological, and biochemical tests. These conventional methods usually produce the right identification of the organism in question; but the answer is not available until the day after receipt of a sample in the laboratory, at the earliest. To speed up identification of microorganisms, efforts have been directed toward the automation of repetitive techniques used in the microbiology laboratory, such as plating, pipetting, making dilutions, and using preformed enzymes and selective growth media. Other approaches to the achievement of rapid automated identification of microorganisms have also been used, including immunologic techniques (1), electrophoresis (2), radioisotopes (3), bioluminscent reactions (4), and pyrolosis gas-liquid chromatography (5).

In previous studies, we reported that gas chromatographic (GC) methods were sensitive and precise enough to detect certain chemical changes in animal body fluids or tissues or in *in vitro* cultures, which were related to characteristic metabolic activities of microorganisms (6–8). Our approach to the problem of rapid detection and identification of microorganisms is based on the assumption that each type of microbial species forms characteristic or possibly unique compound(s) in its growth environment or may yield typical GC elution patterns on analysis of the infected sample. This Chapter deals with feasibility of applications of GC procedures for rapid automated identification of certain microorganisms in clinical specimens.

METHODS AND MATERIALS

Organisms

Bacteria

Bacteria used in these studies were isolated from clinical specimens by plating on human blood agar plates incubated at 37°C for 48 hours. Identification of the bacterial isolates was made by standard diagnostic procedures. A number of bacterial strains were obtained from the American Type Culture Collection (ATCC). Bacterial strains were selected from gram-positive and gram-negative groups frequently encountered in a diagnostic bacteriology laboratory.

Viruses

A mouse neurotropic strain of Influenza A_0 (Wilson-Smith strain) was used for all influenza work. The Indiana serotype of vesicular stomatitis virus (VSV) and a herpesvirus (B) strain were employed for these studies.

Cultural Procedures

Bacteria were initially propagated on appropriate media under aerobic or anaerobic conditions as required by the individual species. For animal innoculations or for GC analysis of *in vitro* cultures, the organisms were grown in Trypticase soy broth (TSB) or brain-heart infusion (BHI) and incubated at 35–38°C for 24 hours. The culture tubes were then centrifuged at 2500 rpm for 10 minutes, and the cells were separated, washed with sterile saline solution, and resuspended in sterile distilled water to the desired dilutions. Mycobacteria were grown on Dubos broth supplemented with pyruvic acid and incubated at 37°C for 2 weeks under a 10% CO_2 atmosphere.

Two to five isolated colonies of bacteria isolated from clinical specimens were transferred to 2 ml of TSB medium supplemented with 3% glucose and incubated at 37°C for 3 hours. Samples were taken from the bacterial cultures for microscopic examination and for plating to determine the viability of the organisms. Replicates of four tubes were used for each species of bacteria analyzed by GC. The culture samples were frozen on dry-ice–alcohol bath and stored at $-20°C$ until they were used for GC analysis.

Influenza virus was grown by dilute passage in 10-day embryonated chicken egg. The VSV strain was cloned and passed at high dilution on BHK_{21} cells. One pool of B virions was used throughout this study. Herpesvirus was isolated by inoculation of tissue cultures of the chorioallantois of the 12-day old chick embroyo.

Animal Inoculations

Rats

White, male Sprague-Dawley (Charles River Breeding Laboratories, Wilmington, Mass.) rats weighing 200 to 300 g were used for GC studies of bacterial infections. Animals were inoculated intraperitoneally with 0.5 ml of washed cell suspension of 10^7 bacterial cells. Uninoculated animals were used as controls with each experiment.

Mice

Young, adult, white Swiss mice of both sexes were used in these studies, for experimental viral infections. Intranasal inoculations of approximately 0.03 ml of virus-containing fluid were carried out under light ether anesthesia. Virus contents of tissues from various organs were determined by plaque assay after the tissues were homogenized in a Virtis tissue homogenizer in 5 ml of Eagle medium.

Specimen Collection

Blood

Experimental animals were bled by cardiac puncture and 2-ml specimens were placed in prechilled vacutainers. Blood samples from infected and uninfected rats were obtained in each experiment. Samples from four mice were pooled for each type of virus studied.

Blood samples (2 ml) from patients and apparently healthy individuals on the hospital staff were obtained in sterile tubes. The samples from patients with known clinical histories and completed bacteriological workups were selected in these studies. All samples from the patients were collected 1 to 4 hours after admission, with no known medication administered to these patients prior to drawing the blood samples. Blood samples from patients with penumococcal pneumonia were collected by Dr. Robert Austrian at the University of Pennsylvania hospital and by Dr. Ronald S. Kundargi at the Hospital of St. Raphael, New Haven, Connecticut. The samples from patients with tuberculosis (TB) were provided by Laurel Heights Sanitarium, Shelton, Connecticut.

Whole blood samples were centrifuged at 4°C for 15 minutes for the separation of serum, and 1 ml of serum was stored at $-20°C$ until the GC analysis was performed.

Tissues

Various tissue samples from infected and uninfected animals were obtained for GC analysis. The animals were anesthetized by ether; and heart, lungs, liver, spleen, kidneys, and brain tissues were quickly removed. The whole organs were thoroughly washed with cold normal saline solution, and 10 to -50 mg tissue samples were quick frozen by using liquid nitrogen. The tissue pieces were stored at $-60°C$ in plastic vials until used for GC analysis.

Sample Preparation

Blood

Whole blood was centrifuged at 4°C for the separation of serum. Sera were prepared for GC analysis as described in Fig. 1. Also, sera from infected patients or samples of bacterial cultures were passed through ion exchange columns to remove interfering ions (9).

Bacterial Cultures

Bacterial isolates subcultured in TSB medium were prepared for GC analysis of Krebs cycle intermediates and related compounds (Fig. 2).

Tissues

Tissue samples were prepared as described in Fig. 3 for the analysis of glycoprotein components and phosphorylated sugars.

Gas Chromatography

Equipment

Two gas chromatographic systems with the capacity of using simultaneously up to four detectors and three columns were employed in these studies. One system was composed of a Varian Aerograph (model 1700) equipped with a flame ionization detector (FID) and an electron capture detector (ECD). A single column was maintained isothermally in this equipment. A splitter (50:50) was used at the end of the column for distributing equal amounts of effluents to the two detectors. The apparatus was fitted with a dual pen recorder (Varian model 20) operating at a chart speed of 40 in./hour. An on-line integrator (Varian model 477) was used to give the printouts of peak areas, retention times (Rt), and baseline correction values.

The second GC system used in these studies consisted of a Hewlett-Packard gas chromatograph (model 1700), dual flame ionization detectors, a temperature programmer (model 240), a dual pen recorder (model 17503A), and a four-channel electronic integrator (model 3370B).

GC Conditions

All GC analyses were performed using 1.8-m long (3-cm inner diameter), stainless steel, coiled columns. Nitrogen was the carrier gas, with flow rate of 40 ml/hour. The detector and injector temperatures were set 10 to 20°C

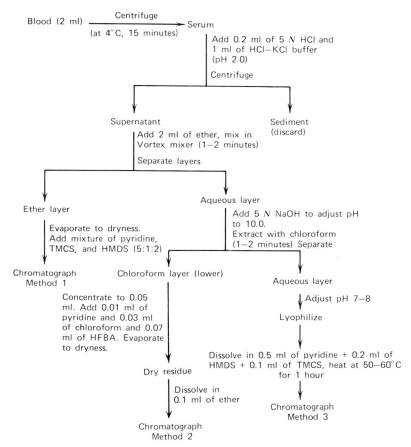

Fig. 1. Preparation of blood sample for gas chromatographic analysis.

higher than column operating temperatures. Other GC conditions varied according to the type of analysis.

Method 1

Acidic and neutral components in samples were separated by the methods described previously (6). Other GC conditions were as follows. Columns: (*a*) 10% Carbowax (20 *M*) on Chromosorb W (60/80 mesh), (*b*) 3% SE-30 on Chromosorb P (80/100 mesh); temperature programmed (TP) from 90 to 225°C at the rate of 5°C/minute; detector, FID and ECD.

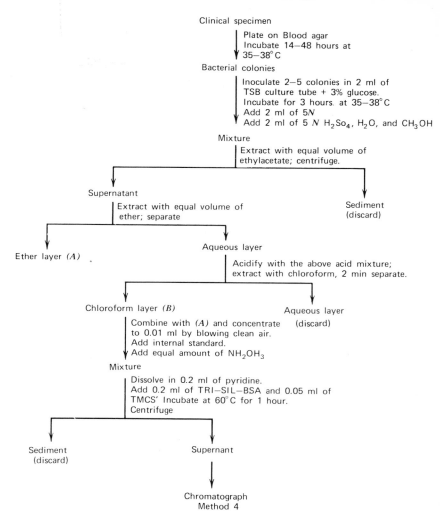

Fig. 2. Preparation of bacterial culture for gas chromatographic analysis.

Method 2

Primary amines and related compounds were analyzed by the methods described by Brooks et al. (10) using columns: (*a*) 10% Carbowax on Chromosorb (60/80) mesh), (*b*) 3% OV-1 on Chromosorb W (80/100 mesh); TP from 80 to 230°C (4°C/minute); detector, ECD and FID.

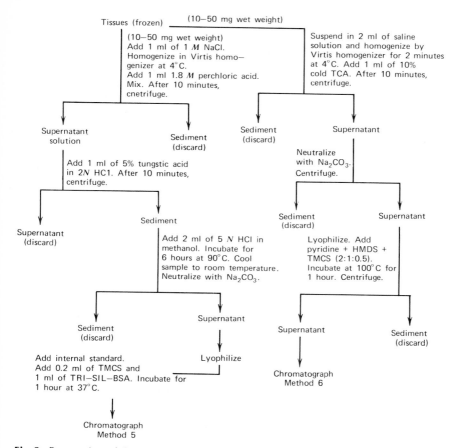

Fig. 3. Preparation of tissue sample for gas chromatographic analysis.

Method 3

Neutral compounds and components of high-molecular-weight compounds were analyzed using columns: (a) 15% EGS on Gas Chrom P (80/100 mesh), (b) 3% SE-52 on Gas Chrom P (80/100 mesh); TP from 110 to 190°C (10°C/minute); detector, ECD and FID.

Method 4

Krebs cycle acids and related compounds produced in TSB cultures by bacterial isolates were analyzed according to a modified method of DeSilva (11) by using columns: (a) 10% PEGA on Chromosorb P (100/120 mesh), (*b*) 5% SE-30 on Chromosorb W (80/100 mesh); TP from 80 to 200°C (5°C/min.); detector, FID.

Method 5

Glycoprotein components in tissues were analyzed according to modified methods of Bolton et al. (12) by using columns: (*a*) 3% SE-30 on Chromosorb W (80/100 mesh), (*b*) 10% Carbowax (20 *M*) on Chromosorb W (60/80 mesh); TP from 90 to 230°C (4°C/minute); detector, FID.

Method 6

Phosphorylated sugars in tissues were analyzed according to methods described by Hashizume and Sasaki (13) with modifications including columns: (*a*) 4% OV-17 on Gas Chrom Q (100/120 mesh), (*b*) 5% silicone fluid (DC-430); TP from 90 to 220°C (2°C/minute); detector, alkali flame ionization.

Data Analysis

Gas chromatographic patterns were compared manually or by using IBM 1130 computer on the basis of:

1. Presence or absence of peaks.
2. Average retention time values ($\pm$ standard deviation); calculated by analyzing 10 replicate samples.
3. Size of peaks (area calculated by on-line integrators).
4. Identification of certain compounds by comparing with the relative retention time (*RRT*) values of standard chemicals.

The data obtained by using various columns, detectors, and extraction procedures were combined, and the common peaks found in uninfected and infected clinical specimens were discarded. The characteristic peaks were then given a number by dividing the chromatograms into specific time intervals (e.g., 20 sec or 30 sec). A second set of numbers as given to the area of the peak [e.g., 00 for peak size $0 < 100$ mm^2 (trace compounds), 01 for peak $>100 < 500$ mm^2, 02 for peak $>500 < 1000$ mm^2]. Identification of organisms was achieved by comparing the data (time intervals and size of the peaks) stored in a computer. The selection of characteristic metabolic products of the organism or host responses due to an organism was based on one or more of the following criteria: (*a*) GC detection of the compound(s) consistently present in the spent culture media and absent from the uninfected control clinical samples or uninoculated media, (*b*) the presence of a compound that would give a peak area of 100 mm^2 or more with the specifically defined GC conditions, (c) the presence of compounds(s) produced by one type of microbial species but not by other organisms, and (d) the presence of compounds (peak) in clinical specimens from different patients (based on statistical analysis of a sample size of 100 patients).

Preliminary attempts were also made to record the GC patterns on a magnetic tape, employing an analog-digital converter or timer to use computer matching (Fig. 4). However, these attempts did not always give satisfactory analyses.

RESULTS AND DISCUSSION

Microbial Metabolite Detection in Clinical Specimens

To apply automated gas chromatographic methods for rapid identification of microorganisms, certain metabolic products generated by the infective agents in clinical specimens were characterized. By GC analysis of sera from experimental rats infected with various bacterial species, at least two peaks were obtained in the chromatograms specifically associated with the particular organism examined (Table 1). Characteristic peaks were detected for each of 50 different bacterial species, representing 20 genera of the organisms studied; a representative sample of the data are presented in Table 1. Peaks were considered to be characteristic when they were present with one type of bacterial infection only and absent from others. The chemical identity of these peaks was not determined. However, the extraction procedures and GC conditions used for the detection of these differentiating peaks indicate that the peaks may be due to short-chain fatty acids, alcohols, ketones, or primary amines. Also, most of the peaks were present in *in vitro* cultures analyzed under similar GC conditions, thereby suggesting specific association of the compounds with metabolic activity of the organisms.

Identification of bacteria based on their characteristic products was further studied by the analysis of glycoprotein components in liver samples of rats infected with bacteria (Fig. 5). There were five to eight compounds detected in these samples. The peaks of *Rt* values 180 and 300 sec were present in uninfected and infected specimens. However, there were some unidentified peaks that were found to be characteristically associated with the activity of a particular infection. For example, liver samples of *D. pneumoniae* infected rats showed peaks of *Rt* values of 360, 630, 1380, 2460, 2820, and 3270 sec whereas in *S. typhimurium* infections, the compounds with

136

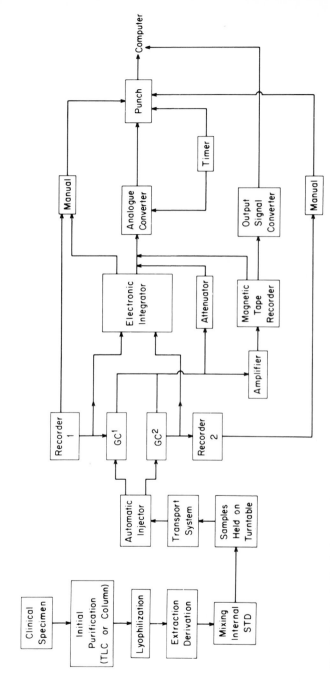

Fig. 4. Schematic block diagram of gas chromatographic system for analysis of clinical specimens.

Table 1 Analysis of Sera from Laboratory Rats Infected with Bacteria[a]

Organism	Characteristic Peaks,[b] Rt (sec)
Diplococcus pneumoniae (type 3)	190, 433, 1114, 2340, 2620
Staphylococcus aureus	620, 915, 2955
Streptococcus (sp. β – haemolytic)	460, 1952, 2305
Salmonella typhimurium	870, 2070, 3450
Pseudomonas aeruginosa	510, 780, 1350, 2100
Haemophilus influenzae	500, 650, 730, 1420, 1710
Klebsiella pneumoniae	180, 360, 750, 1200, 1860, 2940, 3300
Escherichia coli	344, 530, 984, 1262, 1750
Proteus mirabilis	410, 460, 520, 570, 890, 990, 1500
Serratia marcescens	850, 960, 1190, 1320

[a]One ml of serum from 48-hour postinfected animals as analyzed by methods 1 and 2 described in "Methods and Materials."
[b]Peaks were selected on the bases of size (>100 mm^2 area), consistency, and reproducibility. These peaks were also present in *in vitro* cultures. Values represent an average of 10 samples from different animals.

Rt values of 1050, 2280, and 2310 sec were detected in liver specimens. Although the chemical identification of most of these peaks was not achieved, the peaks may be due to pentoses, hexoses, hexoseamines, amino acids, and polysaccharide compounds. However, at least one identified peak was present in the samples from *in vitro* and *in vivo* studies. For example, associated with *D. pneumoniae* infection were erythrose (360 sec) and ribose (630 sec); *S. typhimurium*, mannose (1050 sec); *S. marcescens*, deoxyribose (660 sec); *H. influenzae*; rhamnose (690 sec), xylose (900 sec), glucosamine (1620 sec); *S. faecalis*, sorbose (870 sec), galactitol (990 sec), ribose phosphate (1500 sec), glucosamine (1620 sec) and sucrose (1800 sec); and *S. aureus*, arabinose (780 sec). These studies on glycoprotein analysis of tissue correlate with our previous findings, in which at least five glycoprotein components in sera were characteristically associated with *D. pneumoniae* activities (14).

Sera from human patients with pneumonoccal pneumonia were analyzed for the carbohydrate contents (Fig. 6). Quantitative differences in carbohydrate compounds were observed in sera from bacteremic and nonbacteremic patients when analyzed by method 3. However, the patterns of sera from patients infected with one serotype of pneumococci were different

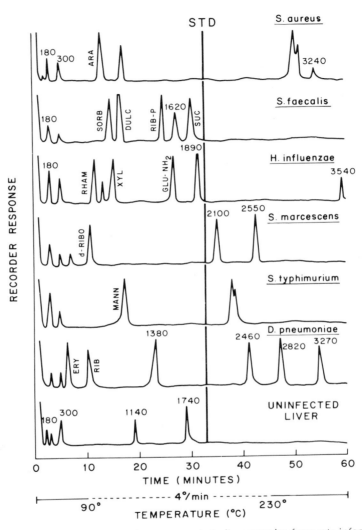

Fig. 5. GC analysis of glycoprotein components in liver samples from rats infected with bacteria. The numerals on the top indicate the average retention time (*Rt*, sec) of determinations from tissues of 10 rats.

from those of the other serotypes. Some of the peaks due to serotype of pneumococci were identified: phosphoglyceric acid (390 sec), fucose (705 sec), galactitol (990 sec), maltose (1740 sec), and fructose-6-phosphate (2280 sec) for type I infections; glycerol-phosphate (300 sec), fucose (705 sec), and galactitol (990 sec) for type 3; glucose (705 sec), galactitol (990 sec), mannitol (1530 sec), and mannose-6-phosphate (2640 sec) for type 4; deoxyribose (660 sec), fucose (705 sec), ribose (840 sec), and xylose (900 sec) for type 6, and α-glycerolphosphate (285 sec), phosphoglyceric acid (390 sec), deoxyribose (660 sec), arabinose (765 sec), α-glucose (1140 sec), and ribose-5-phosphate (1500 sec) for type 8 infection. Other unidentified peaks were also characteristic of pneumococcal types (see Fig. 6). These results suggest that by using ultrasensitive GC methods not only can the bacterial infection be differentiated with regard to the species in a genus, but to some extent the strains within a species can also be identified. These techniques are especially suitable for identification of pneumococci serotypes, since the capsular polysaccharide (specific soluble substance) is known to be immunologically and biochemically distinct for each of the more than 85 types (15).

The GC methods of rapid identification were then applied to the characterization of phosphorylated components in uninfected and virus-infected tissues of mice (Fig. 7). Alterations in phosphorylated sugars were observed in liver, spleen, and brain samples of HV-infected animals, whereas some quantitative changes were detected in kidney, lung, and heart tissues of infected animals when compared with those of uninfected animals. Although efforts to identify the characteristic metabolites in various tissues were not successful, there were 6 phosphorylated sugars in liver samples with Rt values of 540, 840, 1560, 1620, 2580 and 2880 sec which were found to be associated with the activities of herpesvirus. Spleen samples from the infected animals had different peaks (480, 780, 2100, 2940, and 3480 sec), kidney (180 and 1320 sec), heart (180 and 2640 sec), lung (180 sec), and brain (180, 900, 1920, 2400 and 4080 sec) when compared with their respective uninfected tissue specimens. It is interesting to note that specific phosphorylated sugars and nucleic acid components have been observed in certain types of mammary tumors in which a virus etiology is suspected (B. M. Mitruka and W. U. Gardner, unpublished data).

These results show that highly sensitive GC methods can be employed for identification of microorganisms by analysis of specific components that may be generated either by microbial activity *in vivo* or owing to the specific host responses to a particular infectious agent. Identification can be facilitated if the early microbial product detected is a unique metabolite. Our previous studies on viruses (16, 17) and bacteria (8, 18) seem to provide evidence in support of this point of view.

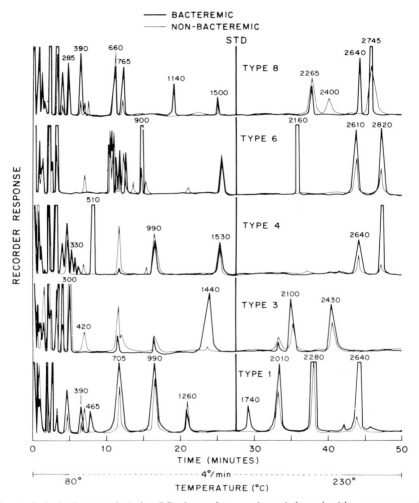

Fig. 6. Carbohydrate analysis by GC of sera from patients infected with pneumococci. The numbers indicate the average *Rt* (sec) values of determinations from samples of three to five patients.

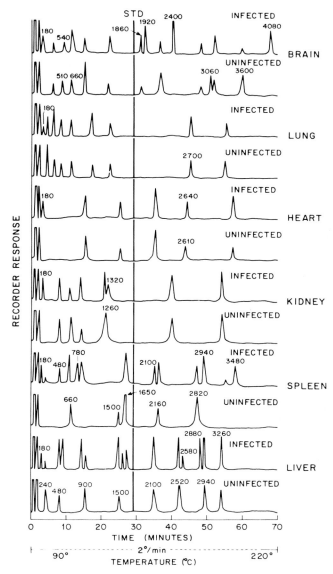

Fig. 7. GC analysis of phosphorylated sugar compounds in tissues of mice infected with herpesvirus B. The numbers indicate the average Rt (sec) values of determinations from 20 animals.

Gas Chromatographic Profiles

Comparison of gas chromatographic patterns of infected and uninfected specimens was used as a means of rapid identification of microorganisms or an infectious disease. A characteristic profile was obtained with each type of bacterial species by the GC analysis of infected sera using methods similar to those just described (Fig. 8). These patterns represent "average" data from the analysis of sera from 4 different patients. The patterns were highly reproducible and easily recognizable, in spite of some variations due to pathophysiological conditions of the individual patients. The presence or absence of each peak in profile was compared by statistical analysis of the sera of 10 to 20 patients, using similar GC methods. A set of consistently occurring peaks in the chromatograms was selected to construct a profile for the specific identification of microorganisms. The reproducibility of GC patterns was evidenced by the comparison of serum analyses from 7 patients infected with *E. coli* (Table 2). Nine characteristic peaks were found to be associated with *in vivo* activities of the organism. Although the areas of these peaks varied, the retention values of each peak had a coefficient of variation (CV) of less than 3%, indicating the similarities of the compounds present in these patients' sera. Also, some other peaks in GC patterns were occasionally present in some of the serum samples but absent from others. Four to six peaks in the patterns had retention values similar to those of uninfected normal controls. These peaks in the chromatographic profiles were not considered to be characteristically associated with the activity of *E. coli*.

Carbohydrate analysis of sera from mycobacterium-infected patients revealed a characteristic and readily discernible profile for this organism (Fig. 9). Representative data comparing the GC patterns of sera from 5 patients with tuberculosis show general similarity, although certain differences in individual patterns are apparent. Many of the peaks detected in sera of patients were also found in the *in vitro* mycobacterial cultures, indicating specific association of these compounds with the activity of the organisms. Using similar methods, chromatograms of 200 sera from TB patients were recorded on magnetic tape and matched by computer (IBM 1130) with those of coded sera of patients with pneumococcal, enteric pathogenic bacilli, and klebsiella infections. The mycobacterial-infected sera were readily identified with this method as well as by manual comparison. Further studies of GC profile comparison as a means of rapid identification of microorganisms were made by analyzing serum samples from virus-infected mice. Figure 10 compares GC patterns of vesicular stomatic virus (VSV), influenza virus (IV), and herpesvirus (HV) infected sera and with those of uninfected. It is apparent that the GC profiles of one type of virus is different from those of the others. The patterns were highly reproducible, as

Table 2 Analysis of Serum Samples from Patients with Pyelonephritis Associated with *E. coli*

Patient	Age (Years)	Bacteremia[b]	Characteristic Peak,[a] Rt (sec)								
			1	2	3	4	5	6	7	8	9
F.L.	28	–	345	540	780	1020	1290	1500	1800	2505	2745
S.A.	21	–	342	540	780	1005	1260	1500	1800	2485	2700
M.C.	31	+	345	525	765	975	1260	1495	1780	2480	2700
F.B.	40	+	336	525	750	975	1260	1490	1740	2490	2750
N.F.	52	–	340	540	765	990	1260	1485	1750	2490	2750
E.W.	25	+	354	525	750	960	1275	1490	1780	2480	2750
M.M.	17	–	345	525	760	960	1280	1500	1780	2485	2750
Av Rt (sec)			343	531	762	984	1269	1495	1776	2475	2735
SD			5.7	8.2	12.6	26.3	12.1	6.2	18.9	16.3	23.6
CV(%)			1.6	1.5	1.6	2.7	1.0	0.4	1.1	0.7	0.9

[a] Selected by the analysis of replicate samples using methods 1 to 3.
[b] Av = average, SD = standard deviation, CV = coefficient of variation (SD/Av × 100).

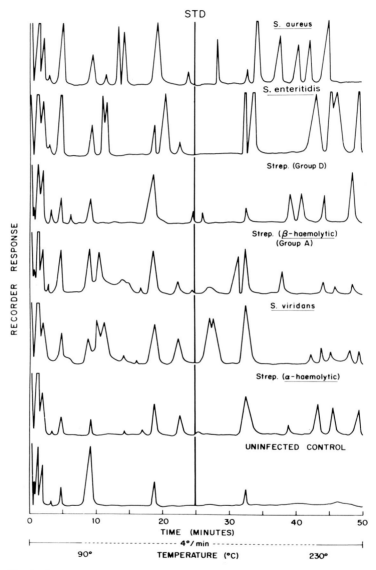

Fig. 8. GC patterns by analysis of sera from patients infected with bacteria.

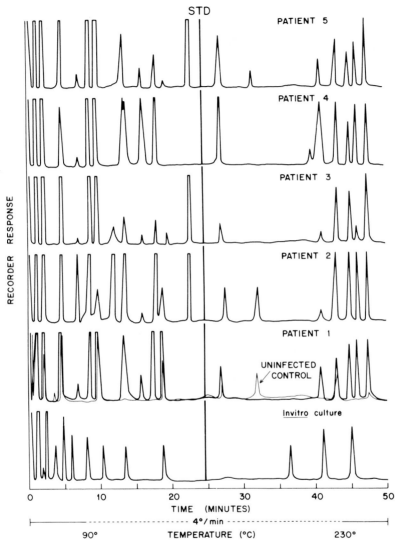

Fig. 9. Comparison of GC patterns by analysis of sera from uninfected and infected patients and *in vitro* cultures of *Mycobacterium tuberculosis*.

145

indicated by the analysis of replicate samples, and they were characteristic of each type of virus examined. These findings support our previous work on the study of certain canine viruses (16) and the equine infectious anemia (EIA) agent (17). Comparison of GC profiles of human clinical specimens in conjunction with mass spectrometry and computer analysis has been reported previously for applications in diagnosis of infectious diseases (18), metabolic disorders (19), and other pathological processes (20, 21).

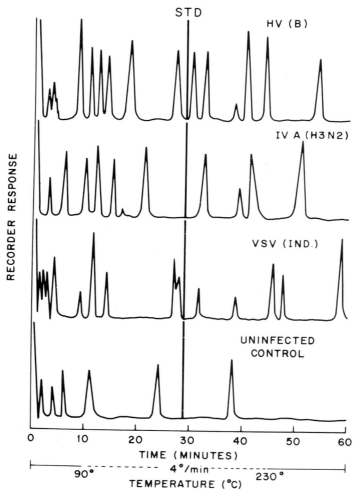

Fig. 10. GC patterns by analysis of sera from mice infected with vesicular stomatitis virus (VSV), influenza virus (IV), or herpesvirus (HV).

Identification of Microorganisms After Primary Isolation from Clinical Specimens

The application of GC methods by direct analysis of clinical specimens for rapid identification of microorganisms is hindered by several factors that may influence the chromatographic patterns—for example, variations of pathogenicity of microorganisms in different hosts, mixed or multiple infections, and effect of therapeutic measures. However, another approach to rapid identification of microorganisms in clinical specimens by GC methods can be used to overcome these problems. In this procedure, microorganisms are isolated from clinical sepcimens on blood agar plates, and the isolated colonies are incubated briefly in an enriched medium to elaborate certain groups of metabolites that are characterized by GC analysis.

A study was made to test the usefulness of this method for rapid identification of microorganisms. Thirty-six strains of bacteria were analyzed for their production of Krebs cycle acids and related compounds. A characteristic pattern of organic acids was obtained for each strain of the test organism (Tables 3–6). Replicate analyses of 10–20 culture tubes containing a single strain of organisms showed remarkably reproducible patterns that were readily discernible. Chromatographic patterns of acids produced by bacteria isolated from coded specimens from patients were compared with the known standards and were accurately identified as *D. pneumoniae*, *S. aureus*, *S. pyogenes*, *S. typhimurium*, *P. vulgaris*, and *K. pneumoniae* species (Fig. 11). The products were essentially similar to those detected with the known bacterial standards (ATCC strains) (Tables 3–6). These results indicate that the identification of unknown bacteria (or other microbial species) in clinical specimens can be achieved in less than 6 hours after primary isolation by examining for the presence or absence of a few typical products in pure cultures.

The application of the GC methods as a routine diagnostic procedure would be possible if the large volume of data were stored in a computer, since the data then could be compared rather rapidly for the identification of unknown microbial species. The data on retention values, areas of chromatographic peaks, identification numbers, and other known characteristics of the organisms were punched on IBM cards; alternatively, the chromatograms were divided into discrete intervals of 20 or 30 sec. An example of the latter method of analysis appears in Table 7. The computer analyses were performed by the comparison of presence or absence of peak(s) in specified intervals and the areas of the peak(s) and other coded characters. Using this procedure, 11 strains of *E. coli* were readily differentiated. The analysis was ordered for selecting that group of peaks which would maximally discriminate among a set of known bacteria. These peaks could

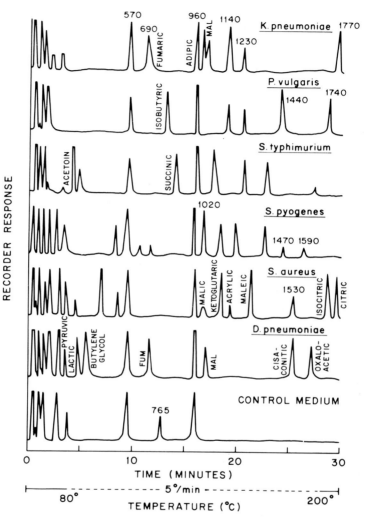

Fig. 11. GC analysis of Krebs cycle acids and related compounds produced by bacteria isolated from patients. Each chromatogram represents the average of six different analyses.

Table 3 GC Analysis of Products in Cultures of Certain Gram-Positive Bacteria[a]

Organisms[b]	Citric	Cisaconitic	Isocitric	Succinic	Fumaric	Malic	Glyoxalic	Oxaloacetic	Ketoglutaric	Lactic	Pyruvic	Acetic	Caproic	Acrylic	Crotonic	Maleic	Diacetyl	Butylene glycol
					Acids												Other	
S. pyogenes (14289)	−	−	+	−	+	−	+	+	−	−	−	−	−	+	+	+	−	−
S. faecalis (19433)	−	−	−	−	−	+	−	−	+	+	−	+	−	−	−	+	+	−
S. aureus (14154)	−	−	+	−	+	+	−	−	−	−	+	−	−	+	−	−	−	−
S. epidermidis (14990)	+	−	−	−	−	+	−	+	−	−	−	−	−	+	−	−	−	−
D. pneumoniae (6317)	−	−	−	−	+	−	+	+	+	+	+	−	+	−	−	+	−	−
D. pneumoniae (6301)	−	+	−	−	+	−	+	+	+	+	−	−	−	−	+	−	−	+

[a]Products were identified by comparison of retention values of standard compounds: +, presence of compound; −, absence of compound.
[b]The ATCC numbers of organisms are given in parentheses.

149

Table 4 GC Analysis of Products in Cultures of *Salmonella* and *Shigella* species

Organism	Citric	Isocitric	Succinic	Fumaric	Malic	Glyoxalic	Oxaloacetic	Ketoglutaric	Lactic	Pyruvic	Formic	Acetic	Acrylic	Crotonic	Mandelic	Maleic	Benzoic	Gb-hydroxybutyric	Acetoin	Diacetyl
S. typhimurium (13311)	−	−	+	−	+	+	+	−	+	−	−	−	−	−	−	−	+	−	+	−
S. typhimurium (14028)	−	−	+	−	−	+	−	+	−	+	−	−	+	−	−	−	−	−	+	−
S. flexneri (12661)	−	+	−	+	−	−	−	−	−	−	+	+	−	−	−	+	−	+	+	+
S. soneii (11060)	+	−	−	−	−	−	−	−	+	−	−	+	+	+	+	−	−	−	−	+

Table 5 GC Analysis of Products in Cultures of Certain Gram-Negative Bacilli

Organism	Acids																	Other		
	Cisaconitic	Isocitric	Succinic	Maleic	Glyoxalic	Oxaloacetic	Ketoglutaric	Lactic	Pyruvic	Acetic	Isobutyric	Caproic	Acrylic	Crotonic	Benzoic	Mandelic	Maleic	Acetoin	Diacetyl	Butylene glycol
Escherichia coli (AB 253)	−	−	+	−	−	−	−	+	−	+	−	−	−	+	−	+	−	+	+	−
Klebsiella pneumoniae (13883)	−	+	−	+	+	−	−	−	+	−	−	±	+	−	+	−	+	−	−	+
Alcaligenes faecalis (19018)	+	−	+	−	−	+	−	−	+	+	−	−	+	−	−	−	−	−	+	−
Proteus mirabilis (14153)	+	+	−	−	−	−	+	+	−	−	−	−	−	+	−	−	+	−	−	−
Proteus vulgaris (13315)	−	+	−	−	−	−	−	−	−	−	+	−	+	+	+	+	−	−	−	−
Pseudomonas aeruginosa (10145)	+	−	−	+	−	−	+	+	−	−	−	−	−	+	−	+	+	−	−	−

151

Table 6 GC Analysis of Products in Cultures of Certain Fastidious Organisms

Organism	Acids																Other		
	Glyoxalic	Pyruvic	Formic	Acetic	Isobutyric	Valeric	Isovaleric	Caproic	Isocaproic	Myristic	Gb-hydroxybutyric	Crotonic	Benzoic	Mandelic	Malic	Succinic	Acetoin	Diacetyl	Butylene glycol
Listeria monocytogenes (15313)	−	+	−	−	−	−	−	−	−	+	−	+	−	+	−	−	−	+	+
Corynebacterium diphtheriae (19409)	−	−	+	−	+	−	−	+	−	+	−	+	+	−	−	+	−	−	−
Haemophilus influenzae (19418)	+	−	−	−	−	+	+	−	+	+	+	−	−	−	+	−	+	−	−
Brucella canis (23365)	−	−	−	+	−	−	−	−	−	−	+	+	+	−	+	−	−	−	−

be treated as characters of bacteria and stored in a computer. For example, out of a maximum of 10 compounds (e.g., organic acids), if each bacterium had a unique set of 5 peaks, 252 combinations can be made; therefore, it should be possible to differentiate 252 bacterial species. However, comparison of an unknown with every chromatographic pattern in the catalog might be time-consuming and costly, even by computer, and if an exact match were not found, there would be no gain from this analysis. Therefore, it would seem to be more reasonable to study the possibility of identifying an unknown first as to genus, then species and strain by the same method. The probability that the unknown is a member of each genus is computed, and it is decided which genus it most likely belongs to, based on a statistical analysis of the data from a set of known bacterial cultures. Current efforts in our laboratory are being directed to using these methods for rapid automated identification of microorganisms in clinical specimens. It might be possible in the near future to apply automated GC methods in a diagnostic microbiology laboratory for identification of organisms either by direct analysis of clinical specimens or by analysis of isolated colonies of organisms grown in suitable culture media.

Table 7 Comparison of GC Data by Computer Analysis[a]

E. coli Strain Number	Peak Interval
T-71	6, 14, 16, 24, 41, 52, 54, 56, 60, 61, 66
K-12	9, 10, 14, 17, 55, 56, 58, 61
6	14, 16, 38, 50, 54, 59
113	4, 5, 6, 14, 31, 32, 66
247	29, 31, 50, 55
253	5, 15, 31, 51
257	34, 38, 45
261	37, 57, 58
262	9, 31, 48, 58, 61, 62, 65
265	35, 40, 51, 62, 65
280	6, 12, 16, 25, 37, 61, 63

[a] Interval = 20 sec; area of peak > 100 mm^2.

Acknowledgments

The author wishes to thank Diana Fischer, Dr. A. J. Stolwijk, Dr. R. S. Kundargi, and Dr. E. Gralla for their interest and helpful suggestions, and M. P. Garcia and B. Su for their technical assistance. This work has been supported by grants from the U.S. Public Health Service (GM 16947 and 2-P06-FR00390).

REFERENCES

1. W. G. Glen, *Aeromed. Rev.*, **I-68**, 1 (1968).

2. V. R. Heubner, *Aerospace Med.*, **1**, 17 (1968).

3. J. K. McClatey, *Inf. Immunol.*, **1**, 421 (1970).

4. G. V. Levin, J. R. Clendenning, E. W. Chappelle, A. H. Hein, and E. Rocek, *Bioscience*, **14**, 37 (1964).

5. E. Reiner, *Nature*, **206**, 1272 (1965).

6. B. M. Mitruka and M. Alexander, *Appl. Microbiol.*, **16**, 636 (1968).

7. B. M. Mitruka, M. Alexander, and L. E. Carmichael, *Science*, **160**, 309 (1968).

8. B. M. Mitruka, A. M. Jonas, and M. Alexander, *Inf. Immunol.*, **2**, 474 (1970).

9. E. C. Horning and M. G. Horning, *Methods Med. Res.*, **12**, 393 (1970).

10. J. B. Brooks, W. B. Cherry, L. Thacker, and C. C. Alley, *J. Infect. Dis.*, **126**, 142 (1972).

11. E. M. de Silva, *Anal. Chem.*, **43**, 1031 (1971).

12. C. H. Bolton, J. R. Clamp, G. Dawson, and L. Hough, *Carbohydr. Res.*, **1**, 333 (1965).

13. T. Hashizume and T. Y. Sasaki, *Anal. Biochem.*, **15**, 346 (1966).

14. B. M. Mitruka, *Yale J. Biol. Med.*, **44**, 253 (1971).

15. R. Austrian, *J. Exp. Med.*, **98**, 21 (1953).

16. B. M. Mitruka, L. E. Carmichael, and M. Alexander, *J. Infect. Dis.*, **119**, 625 (1969).

17. B. M. Mitruka, N. Norcross, and M. Alexander, *Appl. Microbiol.*, **16**, 1093 (1968).

18. E. Jellum, O. Stokke, and L. Eldjarn, *Scand. J. Clin. Lab. Invest.*, **27**, 273 (1971).

19. B. M. Mitruka, R. S. Kundargi, and A. M. Jonas, *Med. Res. Eng.*, **11**, 7 (1972).

20. E. C. Horning and M. G. Horning, *J. Chromatogr. Sci.*, **9**, 129 (1971).

21. A. Zlatkis and H. M. Liebich, *Clin. Chem.*, **17**, 592 (1971).

CHAPTER 9

High-Resolution Field Ionization and Field Desorption Mass Spectrometry of Pyrolysis Products of Complex Organic Materials

H.-R. SCHULTEN

FIELD IONIZATION

Mass spectra obtained by field ionization mass spectrometry (FI–MS) are generally characterized by the presence of prominent molecular ion intensities and minimum fragmentation (1). In principle, therefore, FI–MS is a suitable technique for the analysis of multicomponent mixtures of organic compounds that can be volatilized by evaporation. An example of the potentialities of FI–MS in this area of research was given by Beckey et al. in 1964 (2) in a small comparative study of the qualitative and quantitative analysis of a seven-component hydrocarbon mixture by gas-liquid chromatography (GLC) and by low-resolution FI–MS. It should be pointed out, however, that the interpretation of spectra obtained by low-resolution FI–MS of multicomponent mixtures is inevitably complicated because molecular ions of different elemental composition may share the same nominal m/e value. High-resolution FI–MS therefore greatly widens the scope of the method since it enables the determination of the elemental composition of the observed ions by accurate mass measurements.

In collaboration with the group at the FOM Institute in Amsterdam, where the rapid and reproducible identification of bacteria is performed by fingerprinting them with Curie point pyrolysis GLC (3) and Curie point pyrolysis mass spectrometry (4), our application of high-resolution FI–MS was aimed at the determination of the chemical species that occur in these pyrolysis processes. The purpose of the first qualitative approach was to explore the potentials of high-resolution FI–MS for the analysis of extremely complex multicomponent mixtures. We also wished to perform a general survey of the chemical nature of bacterial pyrolysis products which might serve as a basis for future high-resolution FI–MS studies of biological pyrolyzates.

To this end 5 mg of *Pseudomonas putida* bacteria was pyrolyzed in a simplified pyrolysis procedure (for a detailed description of the experimental conditions and the explicit discussion of the results, see Ref. 5). The volatile products (at 150°C) were examined with high-resolution FI–MS, and more than 100 substances were identified in a single photographically recorded spectrum. As Table 1 indicates, more than 60 of these compounds are identical to those found by Simmonds et al. (6,7) by coupling an *electron impact* (EI) mass spectrometer to the pyrolysis GLC unit. It is rather surprising that the pyrolysis products described by Simmonds match perfectly with our findings, given the differences in the pyrolysis techniques employed and the lack of a close taxonomical relationship between the gram-positive *Bacillus subtilis* and *Micrococcus luteus* strains studied by Simmonds and the gram-negative *Pseudomonas putida* strain analyzed by us.

Table 1 Proposed Chemical Identities of Observed Molecular Ions

Nominal Mass	Elemental Composition	Pryolysis Products: Simmonds (6, 7) and Present Study (5)	Probable Origin[a]	Additional Compounds Found in Present Study (5)
16	CH_4	Methane	P	
17	NH_3	Ammonia	—	
18	H_2O	Water	—	
26	C_2H_2			Ethyne
27	HCN			Hydrocyanic acid
28	CO			Carbon monoxide
28	C_2H_4	Ethene	P, L	
30	CH_2O			Formaldehyde
30	C_2H_6			Ethane
31	CH_3NH_2			Aminomethane
32	O_2			Oxygen
32	CH_3OH	Methanol[b]	N	
33	NH_2OH			Hydroxylamine
34	H_2S	Hydrogen sulfide	—	
35	NH_4OH			Ammonium hydroxide
36	HCl			Hydrochloric acid
41	C_2H_3N	Ethanenitrile	P, N	
42	C_3H_6	Propene	P, L	
44	CO_2	Carbon dioxide		
44	C_2H_4O	Ethylene oxide	P	
44	C_3H_8			Propane
45	CH_3NO			Formamide
46	CH_2O_2			Formic acid
46	C_2H_6O			Ethanol
47	CH_5NO			Hydroxyaminomethane
48	CH_4S	Methanethiol	P	
52	C_2N_2			Cyanogen
53	C_3H_3N	Propenenitrile	P, N	
54	C_3H_2O			Propynal
54	C_4H_6	Butadiene	P, C, L	
55	C_3H_5N	Propanenitrile	P, N	
56	C_3H_4O	Propenal	C, L	
56	C_4H_8	Butene	P, L	
		Methylopropene	P	
58	C_3H_6O	Acetone	C	
		Propanal	C	
58	C_4H_{10}	Butane	?	
59	C_2H_5NO	Acetamide	?	

Table 1 Continued

Nominal Mass	Elemental Composition	Pyrolysis Products: Simmonds (6, 7) and Present Study (5)	Probable Origin[a]	Additional Compounds Found in Present Study (5)
60	$C_2H_4O_2$			Acetic acid
62	C_2H_6S			Ethanethiol or dimethylsulfide
63	HNO_3			Nitric acid
64	SO_2			Sulfur dioxide
67	C_4H_5N	Pyrrole	P, Po	
		Methylpropenenitrile[b]	P	
68	C_4H_4O	Furan	C	
68	C_5H_8	Methylbutadiene	C	
69	C_3H_3NO			Hydroxypropenenitrile
69	C_4H_7N	Butanenitrile	P, L	
		Methylpropanenitrile	P	
70	$C_3H_2O_2$			Propynoic acid
70	C_5H_{10}	Methylbutene	P	
		Pentene[b]	L	
71	C_3H_5NO			Hydroxypropanenitrile
72	C_4H_8O	Butanone	C	
		Methyl propanal	C	
73	C_3H_7NO	Propionamide	?	
74	C_3H_6S	Thiapropane	P	Propenethiol or vinylmethylsulfide
74	$C_3H_6O_2$			Propionic acid or hydroxypropanal
76	C_3H_8S			Propanethiol or ethylmethylsulfide
78	C_2H_6OS			Hydroxyethanethiol
78	C_6H_6	Benzene	P, C	
79	C_5H_5N	Pyridine	P, N	
80	$C_4H_4N_2$	Pyrazine	?	
81	C_5H_7N	Methylpyrroles	P, Po	
82	C_5H_6O	Methylfuran	C	
83	C_5H_9N	Methylbutanenitrile	P	
84	C_6H_{12}	Methylpentene	P, Po	
		Hexene[b]	L	
85	C_4H_7NO			Hydroxybutanenitrile
86	$C_5H_{10}O$	Pentanone	C	
		Methylbutanal	C	
87	C_4H_9NO			Butyramide
88	$C_4H_8O_2$			Butanoic acid or hydroxybutanal

159

Table 1 Continued

Nominal Mass	Elemental Composition	Pryolysis Products: Simmonds (6, 7) and Present Study (5)	Probable Origin[a]	Additional Compounds Found in Present Study (5)
89	$C_4H_{11}NO$			Aminobutanal
90	$C_3H_8O_3$			Glycerol
92	C_7H_8	Toluene	P	
93	C_6H_7N	Methylpyridine	P,N	
94	$C_2H_6S_2$	Dimethyldisulfide[b]	P	
94	C_6H_6O	Phenol	P	
94	$C_5H_6N_2$			Methylpyrazine
95	C_5H_5NO			Hydroxypyridine
95	C_6H_9N	Dimethylpyrroles	Po	
96	$C_5H_4O_2$	Furfural	C	
96	C_6H_8O	Dimethylfuran	C	
97	C_5H_7NO			Furfurylamine
98	$C_5H_6O_2$	Furfuryl alcohol	C	
98	C_7H_{14}	Heptene[b]	L	
100	$C_6H_{12}O$	Methylpentanone	C	
101	$C_5H_{11}NO$			
103	C_7H_5N	Benzenenitrile	P	
104	C_8H_8	Styrene	P	
105	C_7H_7N			
106	C_8H_{10}	Xylenes	P	
		Ethylbenzene	P	
107	C_7H_9N	Dimethylpyridine	P	
108	C_7H_8O	Cresoles	P	
109	$C_7H_{11}N$	C_3-Alkylpyrroles	Po	
110	$C_7H_{10}O$			
111	C_6H_9NO			
112	$C_7H_{12}O$			
112	C_8H_{16}			
113	$C_4H_3NO_3$			
113	$C_6H_{11}NO$			
114	$C_7H_{14}O$			
115	$C_5H_9NO_2$			
116	C_9H_8	Indene[b]	P	
117	C_8H_7N	Indole	P	
		Phenylacetonitrile	P	
		Tolunitrile	P	
118	C_9H_{10}	Methylstyrene	P	
120	C_9H_{12}	C_3-Alkylbenzenes	P	

Table 1 Continued

Nominal Mass	Elemental Composition	Pryolysis Products: Simmonds (6, 7) and Present Study (5)	Probable Origin[a]	Additional Compounds Found in Present Study (5)
122	$C_8H_{10}O$	Ethylphenol	P	
		Xylenoles	P	
123	$C_8H_{13}N$	C_4-Alkylpyrroles	Po	
126	C_9H_{18}	C_9-Alkenes[b]	L, Po	
129	C_9H_7N			
130	$C_7H_{14}O_2$			
131	C_9H_9N	Methylindole		
133	$C_9H_{11}N$			
134	$C_8H_{10}N_2$			
135	$C_4H_9NO_4$			
136	$C_9H_{12}O$			
138	$C_8H_{10}O_2$			
140	$C_9H_{16}O$			
142	$C_8H_{14}O_2$			

[a]As proposed by Simmonds: P, proteins; C, carbohydrates; L, lipids; N, nucleic acids; Po, porphyrins.
[b]Not reported in bacterial pyrolyzates but in pyrolysis studies of organic soil material.

It should be remembered that both experiments are primarily qualitative, and the qualitative similarity of the major chemical building blocks encountered throughout the microbiological world needs no comment. There are two main differences, however, between the information available from the combination of the GLC–EI mass spectrometer and our direct investigations with high-resolution FI–MS.

First, since compounds within a wide range of polarity are produced in the pyrolysis process, it is difficult to obtain ideal gas chromatography conditions. Thus several of the more polar pyrolysis products (e.g., free acids, amines) do not appear to pass through the GC column but are identified by direct evaporation and FI–MS of the mixture (cf. Table 1, columns 3 and 5).

Second, compounds such as branched hydrocarbons, amines, and alcohols are difficult to identify in GLC–MS studies because they often fail to produce stable molecular ions on electron impact ionization. The softer field ionization process, on the other hand, yields almost exclusively molecular ions or quasi-molecular ions from all pyrolysis products.

FIELD DESORPTION

A new technique, pyrolysis field desorption mass spectrometry (Py-FD–MS) has been developed for the analysis of biological material. This technique involves direct pyrolysis of submicrogram quantities (5×10^{-8} g) of a biological sample on the high-temperature-activated (8) tungsten wire emitter in the field desorption source of a high-resolution mass spectrometer. A combination of favorable conditions for the observation of large pyrolysis products is obtained by pyrolyzing directly in the extremely high field ($>0,1$ V/$\overset{\circ}{A}$) on the surface of an activated emitter. These conditions are:

1. Minimal sample quantities (10^{-8}–10^{-9} g).
2. Minimal difference in time and place between pyrolysis and ionization.
3. Minimal excess energy (0.2 eV) transferred during ionization (1).
4. Minimal residence time (10^{-11} sec) of the ions in the pyrolysis zone (1).

The new technique has been applied to herring DNA; all five bases (i.e. cytosine, methylcytosine, thymine, adenine, and guanine), as well as all nucleosides, nucleotides, and some dinucleotides, have been identified. The prospects of obtaining base sequence information by Py-FD–MS are discussed in Reference 9. Further applications of Py-FD–MS to the analysis of biological macromolecules such as polypeptides and polysaccharides are being investigated. The studies are being undertaken to resolve some of the reaction mechanisms in Py-FD–MS as well as to explore more fully the potential of the method for the structural analysis of biopolymers. The first step was the Py-FD mass spectrum of glycogen, a branched polymer of amylose chains joined through $\alpha(1\rightarrow6)$ linkages (Fig. 1). Although this polysaccharide is a highly elaborate molecule, the Py-FD–MS in Fig. 2 displays some significant molecular and quasi-molecular ions derived from mono- and disaccharide units. High-resolution data of the peaks at m/e 262, 324, and 340 indicate building blocks of the biopolymer (see Fig. 1).

Other signals are certainly due to compounds generated by pyrolytic bond rupture, water elimination, and subsequent protonation of the dehydrated molecule. The peak at m/e 99.042 ($C_5H_7O_2$), for example, is easily identified as the $(M+H)^+$ peak of furfuryl alcohol or isomers. This identification is proved by precision mass measurements, from our library of pyrolysis products of sugar-containing material; beyond this, the compound has been identified in bacterial pyrolysis (5).

Since the studies mentioned are in the preliminary stage, it is not yet possible to say whether the method has potential for distinguishing between branched and linear polysaccharides. Further investigations will reveal how much other structural information can be obtained.

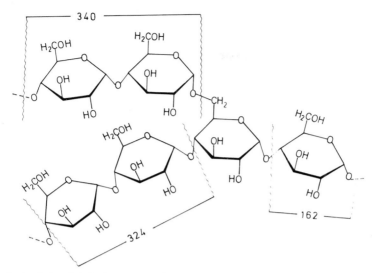

Fig. 1. Part of the $D(+)$-glycogen molecule $(C_6H_{10}O_5)_n$. Molecular weight 270,000 to 3,500,000.

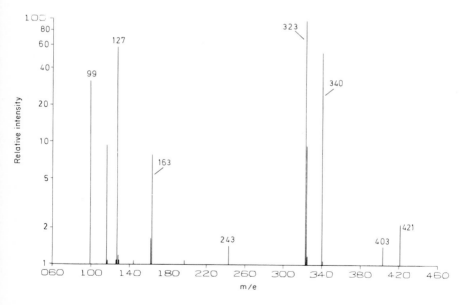

Fig. 2. Pyrolysis field desorption mass spectrum (Py-FD–MS) of glycogen. Photographic detection with vacuum-evaporated AgBr plate (Ionomet, Burlington, Mass.). Approximate resolution: ≈20,000 (10% valley definition).

The application of Py-FD–MS to polysaccharides, polypeptides, and polynucleotides (e.g., to detect structural alterations in the building blocks of DNA due to ionizing radiation or chemical agents, is being studied in our laboratories. This work includes the use of a better defined pyrolysis technique, such as Curie point pyrolysis in direct combination with high-resolution FI–MS, and an automated regulation of the time/temperature profile for Py-FD–MS.

Acknowledgments

The author wishes to thank Professor H. D. Beckey for his continuous support and encouragement.

REFERENCES

1. H. D. Beckey, "Field Ionization Mass Spectrometry," Pergamon Press, Oxford, and Akademie Verlag, Berlin, 1971.

2. H. D. Beckey, H. Knöppel, G. Metzinger, and P. Schulze, in *Advances in Mass Spectrometry*, Vol. 3, W. H. Mead, Ed., The Institute of Petroleum, London, 1966, p. 35.

3. H. L. C. Meuzelaar, B 81, Symposium on Rapid Methods and Automation in Microbiology, Stockholm, June 1973.

4. Piet G. Kistemaker, B 82, Symposium on Rapid Methods and Automation in Microbiology, Stockholm, June 1973.

5. H.-R. Schulten, H. D. Beckey, H. L. C. Meuzelaar, and A. J. H. Boerboom, *Anal. Chem;* **45**, 191 (1973).

6. P. G. Simmonds, *Appl. Microbiol.*, **20**, 567 (1970).

7. P. G. Simmonds, G. P. Shulman, and C. H. Stembridge, *J. Chromatogr. Sci.*, **7** 36 (1967).

8. H.-R. Schulten and H. D. Beckey, *Org. Mass Spectrom.*, **6**, 885 (1972).

9. H.-R. Schulten, H. D. Beckey, A. J. H. Boerboom, and H. L. C. Meuzelaar, *Anal. Chem.*,**45**, 2358 (1973).

CHAPTER 10

Rapid and Automated Identification of Microorganisms by Curie-Point Pyrolysis Techniques

I: Differentiation of Bacterial Strains by Fully Automated Curie Point Pyrolysis Gas-Liquid Chromatography

HENK L. C. MEUZELAAR, PIET G. KISTEMAKER, ANNEKE TOM

INTRODUCTION

Analysis of organic materials, including biological compounds, by means of pyrolysis techniques typically follows the pattern outlined in Fig. 1. The scope of this Chapter is restricted to pyrolysis methods applied to microbiological samples. More general information on pyrolysis techniques can be obtained from excellent articles by Levy (1) and Walker (2).

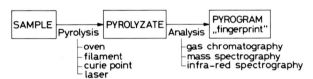

Fig. 1. Alternative pyrolysis methods reported in the literature for the characterization of organic materials.

Pyrolysis in combination with gas-liquid chromatography (Py-GLC) for differentiation and classification of microorganisms was successfully used by Reiner in 1965, following pioneering experiments by Oyama in 1963 (3,4). Especially Reiner continued and extended these studies, and his reports now cover a substantial number of bacterial species and strains (5–8), demonstrating the remarkable chemotaxonomic versatility of Py-GLC. Although the potentials of Py-GLC have been confirmed by other workers (9–13), the technique has not yet evolved into an accepted tool for routine bacterial identification.

Some of the reasons for this lack of acceptance by bacteriologists are easily understood:

1. Interlaboratory reproducibility has not been achieved in Py-GLC of biological samples and will probably never be attained by the oven and filament pyrolysis techniques most often used in this field (14).

2. Py-GLC of complex biological samples, such as microorganisms, is not one of the most accessible GLC techniques, since it requires the use of highly efficient columns and temperature programming. Moreover, operation of Py-GLC systems is often complicated by contamination and/or column degradation problems.

167

3. There is no apparent relationship between the results of classification by conventional techniques (e.g., serology) and those obtained by Py-GLC. This tends to make the latter results umpredictable and inconsistent when judged by the standards of present classification schemes.

For the past few years our group at the FOM Institute has been carrying out a research program specifically aimed at the elimination or alleviation of the above-mentioned problems.

One of the first steps was to select a pyrolysis technique offering prospects for achieving interlaboratory reproducibility. There seemed to be only two candidate techniques; the Curie point pyrolysis method developed by Simon et al. (15) and the capacitor discharge filament technique described by Levy (16). Both techniques are characterized by fast temperature rise times and highly reproducible temperature/time profiles.

The basic parameters of Curie point pyrolysis have been thoroughly investigated (17), and the method is rapidly gaining acceptance in the field of technical polymers (18). Moreover, instruments of this sort are commercially available (17). Preliminary studies showed that the Curie point method could be readily adapted to the analysis of complex biological samples such as bacteria (12). Furthermore, the technique is well suited to automation because the heating wires are disposable and the samples can be prepared in advance.

Since we felt that full automation might make Py-GLC more accessible to laboratories lacking GLC experience, provided adequate maintenance services are at hand, we constructed a completely automated Py-GLC system operating with commercially available capillary GLC columns (19). Finally, studies on isolated fractions of streptococcal cell walls in collaboration with Dr. J. H. J. Huis in 't Veld from Laboratory for Microbiology, State University, Utrecht, demonstrated that biochemical analysis of the cell wall can establish a direct link between serological properties and pyrolysis patterns (20).

A second line of research in our group has been directed toward a completely different approach to obtaining characteristic and reproducible patterns of pyrolysis products—namely, pyrolysis mass spectrometry (Py-MS). Although Py-MS studies of proteins were reported by Zemany (21) as early as 1952, the method does not seem to have found further application in biology, and its development was actively pursued only in the field of technical polymers (22). Detailed descriptions of our Py-MS method, combining Curie point pyrolysis with quadrupole mass spectrometry and signal averaging, can be found elsewhere (23,24). Some recent results obtained with microorganisms are described in Chapter 11 of this volume.

Assuming that the encouraging results obtained by this method will be confirmed by further studies and that Py-MS systems will become commercially available, what are its prospects of becoming an accepted technique in microbiology? In the first place, Py-MS is not plagued by problems analogous to column degradation or differences in column properties, such as we encounter in Py-GLC. This improves the chances of achieving interlaboratory reproducibility. Second, operating a simple Py-MS system of the type described by us probably requires less training and background knowledge than operating a non automated Py-GLC system. Finally, some information on the chemical nature of the sample is directly available from the mass pyrograms (24), and this facilitates the comparison of Py-MS results with other biochemical data.

The price of a Py-MS system, roughly two to four times as much as a comparable Py-GLC system, is strongly compensated by a 10 to 50 times higher analysis rate. In combination with the ease of computer processing of mass pyrograms, this offers attractive possibilities for compiling computer libraries of reference pyrograms. In fact, Py-MS has so many technical and practical advantages over Py-GLC that the future viability of the latter technique in microbiology seems to depend on one all-important question. Which technique affords the most precise differentiation of microorganisms? Our present experience with a limited number of bacterial species, such as *Streptococci, Klebsiella, Vibrio, and Mycobacteria*, indicates that no unequivocal answer can be given. Perhaps 90% of all successful differentiations achieved by combining both techniques could have been made by Py-GLC alone and about 80% by Py-MS alone. Thus the techniques obviously complement each other, with Py-GLC discriminating slightly better than Py-MS. At present, therefore, active development in both directions seems to be well worthwhile.

This Chapter describes a new development in Py-GLC—analysis of streptococcal strains directly taken from blood agar culture plates, by fully automated Curie point pyrolysis gas-liquid chromatography in combination with computer identification of the pyrograms.

EXPERIMENTAL

Samples

Blood agar cultures of streptococcal strains were kindly provided by Dr. J. H. J. Huis in 't Veld, Laboratory for Microbiology, State University, Utrecht. The blood agar plates were incubated at 37°C for 48 hours. Strains of the *Streptococcus mutans* subgroup were incubated in an anaerobic atmosphere (10% $CO_2 + 90\%$ H_2) to facilitate their growth. Strain Z_3 is a mutant of Z_3III, lacking the type III antigen (26). Strains Z_3 and Z_3III were grown under aerobic as well as anaerobic conditions to study possible influences on the pyrolysis patterns.

Pyrolysis Technique

We take either one large colony or a few small ones directly from the culture plates by means of a looped platina wire. About 50 to 100 μg (estimated dry weight) of this material are transferred to the ferromagnetic heating wires used in Curie point pyrolysis by gently rotating the tip of such a wire against the bacterial substance on the looped platina wire. The Curie point wires are then dried in a vacuum chamber at a pressure of 10 torr (T) for 3 to 5 minutes, mounted into a glass reaction tube (see Fig. 2) and subsequently transferred to an automatic sample exchanger (Fig. 3).

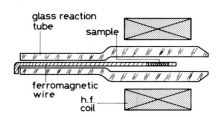

Fig. 2. Basic arrangement for Curie point pyrolysis, as used by the authors in pyrolysis gas-liquid chromatography as well as in pyrolysis-mass spectrometry.

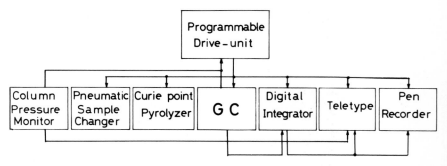

Fig. 3. Schematic representation of the fully automated Curie point pyrolysis gas-liquid chromatography system.

At the appropriate time of the analysis cycle this sample exchanger transports the glass reaction tube with the sample into the Curie point pyrolysis reactor, which is directly coupled to the inlet of the GLC column. Inside the reactor the sample is centered in a high-frequency coil (Fig. 2) connected to a HF power supply (Fischer, 1.1 MHz, 1.5 kW).

The sample is pyrolyzed by energizing the HF coil, which causes rapid heating of the ferromagnetic wire. When the wire reaches the Curie temperature, the temperature of the wire stabilizes until the HF power supply is switched off. A detailed account of the basic principles involved can be found elsewhere (17).

All samples were pyrolyzed on Fe/Ni wires of 0.5 mm diameter, having a Curie temperature of 610°C and a temperature rise time of 100 msec with the Fischer power supply. Total heating time was 1 sec. The pyrolysis reactor is a metal version of a glass prototype described earlier (12) and is specially adapted to automatic operation in combination with the pneumatic sample exchanger. A technical description of this system, constructed at our laboratory, will be published elsewhere (19).

Gas Chromatography System

Figure 3 gives a schematic representation of the automated Py-GLC system. The gas chromatograph, a Becker Delft multigraph (model 409) has a single, 32-m-long, capillary SCOT column (Perkin-Elmer) coated with 10% Carbowax 20 M. The chromatograph is equipped with flame ionization detectors kept at 225°C. The temperature program used appears in Fig. 4. A matrix programming board (Becker "Automatrix") automatically controlls temperature programming, column reconditioning, oven cooling, and restabilization at the initial temperature.

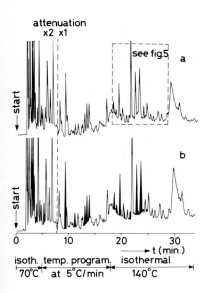

Fig. 4. Pyrograms of replicate cultures of *Streptococcus* Z_3III sampled directly from blood agar plates. (*a*) Aerobic culture, (*b*) anaerobic culture. A typical "fingerprint region" is indicated in pyrogram *a*. (see also Fig. 5). Thirteen peaks selected for computer matching are solid black in *b*. Pyrolysis conditions are given in the text.

Timing signals available at the matrix board enable a specially built electronic drive unit to switch all other units on and off at the appropriate time of the analysis cycle (Fig. 3). These units include a Hewlett Packard electronic digital integrator (model 330 A), a teletype (Teletype, model ASR 33), and a pen recorder.

An electronic pressure transducer with digital readout and printout facilities, continuously monitors column pressure. After sample exchange, this pressure detector blocks all further events until column pressure has returned to normal. Furthermore, it can detect excessive changes in column pressure due to leaks or plugging, in which case the oven will be cooled immediately to protect the column. The duration of a complete analysis cycle can be varied by means of the matrix board, but 45 to 60 minutes is normally needed for effective separation of pyrolysis products, including reconditioning, cooling, sample exchange, and restabilization. The pneumatic sample exchanger has a capacity of 24 samples, and since the system is capable of continuous, unattended operation, a maximum of 24 to 30 samples can be analyzed per 24 hours.

RESULTS AND DISCUSSION

Five replicate cultures of Z_3III and Z_3 strains were analyzed over a period of one month. Growing the cultures under anaerobic conditions or storing the blood agar plates for 5 days at 4°C after the normal incubation period did not noticeably affect the pyrograms. However, since the samples were not washed at all, in contrast with previous studies (12, 20), it was expected that the choice of the culture medium would influence the pyrolysis patterns. Indeed, a moderate to strong influence was observed when the strains were grown on other media (e.g., glucose agar). Nevertheless, these findings seemed to be of little relevance to the present study, since blood agar plates have been quite satisfactory for growing, harvesting, and differentiating all streptococcal strains thus far analyzed. As yet, we have not thoroughly investigated the influence of different media and all results discussed here strictly refer to blood agar cultures. An example of the high degree of qualitative and quantitative reproducibility obtained is given in Fig. 4. The upper pyrogram shows a Z_3III strain grown under aerobic conditions as opposed to the anaerobic culture in the lower pyrogram. Moreover, the pyrograms were obtained on different days.

Mean relative peak height deviation between pyrograms of replicate cultures typically varies from 10 to 15%. It should be noted, however, that these values represent the sum of all variations due to culturing, sampling, pyrolysis, GC analysis, and peak height measurements. Of course, the degree of variability observed in pyrograms of replicate cultures should be compared against the magnitude of the differences observed between nonidentical strains. Depending on the experience of the observer, visual comparison of the pyrograms enables a qualitative or even semiquantitative assessment of these differences. This comparison is facilitated by concentrating on characteristic regions of the pyrograms, sometimes called "fingerprint regions" (6), which may be empirically selected.

In our experience, the region indicated in Fig. 4a is highly characteristic for the series of streptococcal strains analyzed, as illustrated in Fig. 5. This figure demonstrates that some strains, such as sanguis, show relatively large differences, whereas other strains seem to be more closely related. Although visual comparison tends to be complicated by differences in relative sample weight, even strains within the mutans group are easily differentiated.

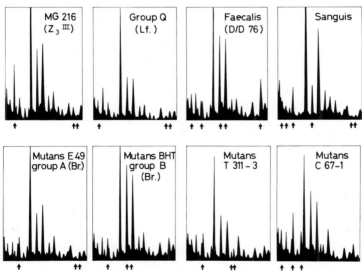

Fig. 5. Fingerprint regions from pyrograms of eight streptococcal strains. The figures have not been corrected for sample weight. Characteristic combinations of relative peak amplitudes are indicated by arrows; (Br.) refers to the subgroups of *Streptococcus mutans* proposed by Bratthåll.

Notwithstanding the obvious success of visual differentiation, the need for more quantitative and objective methods for the assessment of pyrograms has been recognized by most workers in this field. Menger et al. reported successful computer matching of nine out of ten *Salmonella* strains, using peak height values of up to nine characteristic peaks in the pyrograms (26). Our approach is similar in some respects but differs in that it provides a calculated degree of correspondence between each pair of pyrograms instead of giving a "yes" or "no" response. To calculate this degree of correspondence,as many as 30 peaks were initially measured in each pyrogram.

Such measurements may be performed by means of the digital integrator indicated in Fig. 3, thereby yielding a punched tape record of peak areas and corresponding retention times. The tapes can be processed by computer, using a computer program that recognizes the peaks by matching the retention times against a file of retention time data on several different bacteria. Test runs showed that the peaks could indeed be reliably recognized, but the integrator made serious errors in calculating peak areas of incompletely resolved peaks. We therefore decided to use hand-measured peak height values for the studies reported here.

The measured peak heights were normalized with respect to sample weight by expressing them as percentage of total peak height. Subsequently, the variation observed in different streptococcal strains, relative to the variation in replicate cultures was examined for each peak. Thirteen peaks showing the best ratio between these two types of variation were selected for computer matching of all unknown pyrograms with a file of single pyrograms of ten different streptococcal strains. These peaks are indicated in Fig. 4b.

The computer program compares each peak height value of the unknown pyrogram with the corresponding value of the filed strain, always dividing the smaller value by the larger one. Thus 13 peak height ratios are obtained which are smaller or equal to 1. In the case of two completely identical pyrograms, all ratios will of course be equal to 1, whereas the ratios of two randomly differing pyrograms will be scattered between 0 and 1.

In practice, no completely identical pyrograms are encountered; therefore, the peak height ratios of replicate cultures may vary between .85 and 1.00, in line with the aforementioned level of quantitative reproducibility. Figures 6, 7, and 8 show some examples of the results of the computer matching procedure. To obtain a single parameter for the degree of correspondence between two pyrograms, the arithmetic mean of the 13 peak height ratios was calculated. The correspondence values of three unknown cultures, later disclosed to be replicate cultures of Z_3III, with the file of 10 known strains (Fig. 6) demonstrate that a correspondence degree higher than .85 is only observed for the filed Z_3III strain. The same correct results were obtained with replicate cultures of the mutant Z_3, lacking the type III polysaccharide antigen, as shown in Fig. 7.

Two other features of the bar graphs in Figs. 6 and 7 are also apparent. In the first place, the bar graph profiles in Fig. 6 are remarkably similar to one another, and the same applies to Fig. 7. This strongly indicates that the statistical procedure used to obtain the correspondence values gives reliable

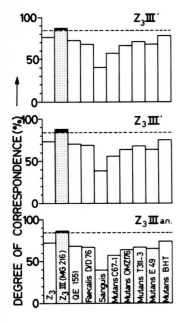

Fig. 6. Computer-calculated correspondence degrees for pyrograms of three replicate cultures of Z_3III with the filed pyrograms of 10 different streptococcal strains. These strains are identified by name in the lower bar graph only; "Z_3III an." was grown under anaerobic conditions. Note that only identical strains have a correspondence degree higher than 84%.

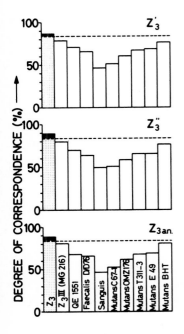

Fig. 7. Computer-calculated correspondence degrees for three replicate cultures of Z_3 with the filed pyrograms of 10 different strains. See also Fig. 6.

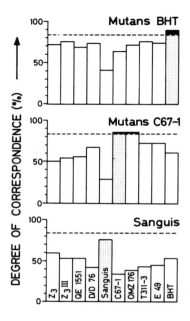

Fig. 8. Computer-calculated correspondence degrees for pyrograms of three different strains with the pyrograms of the 10 filed strains. The names of the filed strains have been abbreviated.

quantitative results. Therefore, a rather sharp line can be drawn between the normal range of variation in replicates and the variation encountered in different strains. In our experience, this line is best drawn at a correspondence degree of 84%, as illustrated in the bar graphs. Second, the figures show that whereas some strains seem to be relatively closely related, *Streptococcus sanguis* differs considerably from all other strains. It will be recalled that similar conclusions were drawn upon visual assessment of the "fingerprint regions" in Fig. 5. Nevertheless, the calculated correspondence degree, although apparently reliable for matching a limited number of streptococcal strains, is not inherently representative for the chemotaxonomic relationships, since it is only based on a selected number of peaks. Moreover, it is extremely unlikely that these complex relationships can be accurately described by a one-dimensional parameter such as the calculated correspondence degree. Obviously, more sophisticated computer classification programs, incorporating multidimensional correlation functions, will have to be used for such studies.

In line with the foregoing considerations, it is not surprising that the present matching procedure using a library of single pyrograms and a single correspondence value also has its limitations, as illustrated in Fig. 8. Although the mutans BHT culture was correctly matched (Fig. 8, upper bar graph), two equivalent matches were found in the case of a mutans C67-1 culture—namely, mutans C67-1 and mutans OMZ176. Visual inspection of the complete fingerprinting patterns (not shown) did not resolve this problem. This raised the interesting question of whether these two strains, representing different isolates of *Streptococcus mutans*, might not in fact be identical. Thus far, this speculation has been borne out by independent Py-MS analysis, but detailed serological studies will be needed to settle this question.

A second type of problem sometimes encountered is also demonstrated in Fig. 8. The lower bar graph shows the correspondence values obtained for an unknown culture, later disclosed to be a sanguis strain. Although the correspondence degree obtained with the filed sanguis strain is markedly higher than with the rest of the file, the degree does not reach the confidence line at 84%. This means either that the unknown was not represented in the file or that the reproducibility of this particular analysis was not satisfactory. Visual comparison of the pyrograms, however, strongly indicated that the unknown was indeed a sanguis strain. Therefore, the analysis was repeated, and this time the culture was correctly identified as *Streptococcus sanguis*.

On the basis of the results discussed in here, we feel that automated Py-GLC in combination with computer evaluation of pyrograms offers a powerful tool for rapid identification of streptococcal strains. Remaining weak links in the system, such as the inadequate performance of the digital integrator with partially resolved peaks, are now being investigated, and the scope of our studies has been extended toward bacterial species of general medical interest such as *Klebsiella and Mycobacteria*.

Acknowledgments

The authors thank the technical staff at the FOM Institute for expert assistance in these investigations. Also they are indebted to Professor Dr. J. Kistemaker and Dr. J. H. J. Huis in 't Veld for constant support and stimulating discussions.

The research was supported by the Organization for Fundamental Research on Matter (FOM) and by the Dutch Ministry of Health.

REFERENCES

1. R. L. Levy, *Chromatogr. Rev.*, **8**, 48 (1966).
2. J. Q. Walker, *Chromatographia*, **5**, 547 (1972).
3. E. Reiner, *Nature*, **206**, 1272 (1965).
4. V. I. Oyama, *Nature*, **200**, 1058 (1963).
5. E. Reiner, *J. Gas Chromatogr.*, **5**, 65 (1967).
6. E. Reiner, J. J. Hicks, R. E. Beam, and H. L. David, *Amer. Rev. Resp. Dis.*, **104**, 656 (1971).
7. E. Reiner, J. J. Hicks, M. M. Ball, and W. J. Martin, *Anal. Chem.*, **44**, 1058 (1972).

8.   E. Reiner and J. J. Hicks, *Chromatographia*, **5**, 525 (1972).

9.   V. I. Omaya and G. C. Carle, *J. Gas Chromatogr.*, **5**, 151 (1967).

10. R. D. Cone and R. V. Lechowich, *Appl. Microbiol.*, **19**, 138 (1970).

11. Ph. G. Vincent and M. M. Kulik, *Appl. Microbiol.*, **20**, 957 (1970).

12. H. L. C. Meuzelaar and R. A. in 't Veld, *J. Chromatogr. Sci.*, **10**, 213 (1972).

13. A. S. Sekhon and J. W. Carmichael, *Can. J. Microbiol.*, **18**, 1593 (1972).

14. R. L. Levy, *J. Gas Chromatogr.*, **5**, 107 (1965).

15. W. Simon and H. Giacobbo, *Angew. Chem. Int. Edit.*, **4**, 938 (1965).

16. R. L. Levy, D. L. Fanter, and C. J. Wolf, *Anal. Chem.*, **44**, 40 (1972).

17. Ch. Bühler and W. Simon, *J. Chromatogr. Sci.*, **8**, 323 (1970).

18. N. B. Coupe, C. E. R. Jones, and P. B. Stockwell, Lecture delivered at the Second International Sympsium on Pyrolysis Gas Chromatography, Paris, 1972.

19. H. L. C. Meuzelaar, H. G. Ficke, and H. C. den Harink, *J. Chromatogr. Sci.* (1974), in press.

20. J. H. J. Huis in 't Veld, H. L. C. Meuzelaar, and A. Tom, *J. Appl. Microbiol.*, July 1973.

21. P. D. Zemany, *Anal. Chem.*, **24**, 1709 (1952).

22. H. D. R. Schüddemage, Thesis, Cologne University, 1967.

23. H. L. C. Meuzelaar and P. G. Kistemaker, *Anal. Chem.*, **45**, 590 (1973).

24. H. L. C. Meuzelaar, M. A. Posthumus, P. G. Kistemaker, and J. Kistemaker, *Anal. Chem.*, **45**, 1546 (1973).

25. J. M. Willers and G. H. J. Alderkamp, *J. Gen. Microbiol.*, **49**, 41 (1967).

26. F. M. Menger, G. A. Epstein, D. A. Goldberg, and E. Reiner, *Anal. Chem.*, **44**, 423 (1972).

CHAPTER 11

Rapid and Automated Identification of Microorganisms by Curie-Point Pyrolysis Techniques

II. Fast Identification of Microbiological Samples by Curie-Point Pyrolysis Mass Spectrometry

PIET G. KISTEMAKER, HENK L. C. MEUZELAAR,
AND MAARTEN A. POSTHUMUS

INTRODUCTION

In comparison with gas chromatography, mass spectrometry has played a minor role in the analysis of bacterial pyrolysis products thus far. A combined gas chromatography–mass spectrometry system was used by Simmonds (1) for the chemical identification of more than 60 pyrolysis products of two bacterial strains. Direct introduction of bacterial pyrolyzates into a high-resolution mass spectrometer equipped with a special ionization source enabled Schulten et al. (2) to determine the elementary composition of more than 100 components, including most of the products identified by Simmonds.

Recently a different application of pyrolysis mass spectrometry (Py-MS) was described by Meuzelaar and Kistemaker (3). By using a Curie point pyrolysis system in combination with a small quadrupole mass spectrometer, reproducible fingerprints of microorganisms could be obtained. Further developments of this system considerably improved the characteristicity of the fingerprints (4).

Moreover, this Py-MS technique has some important advantages over pyrolysis gas chromatography, as discussed in Chapter 10. Therefore, further Py-MS studies of microorganisms were undertaken. We want to report here some of the results, especially with regard to differentiation and computer classification of a limited series of streptococcal strains.

EXPERIMENTAL

Sample Preparation

The sample preparation for the streptococcal strains is described in Chapter 10. Two more strains were analyzed, a *Klebsiella* strain grown on a blood agar plate for 24 hours and a *Vibrio* strain incubated in anaerobic atmosphere ($10\% \ CO_2 + 90\% \ H_2$) for 72 hours.

Apparatus

Since the Py-MS system (Figs. 1 and 2) is described elsewhere in detail (3, 4), we mention here only the main features of the system. The sample is inserted in the apparatus through a vacuum lock and pyrolyzed in a glass reaction tube, directly at the inlet of the expansion chamber. The pyrolysis products drift through a 0.7-mm leak orifice into a quadrupole mass analyzer, where the fragment molecules are ionized and selected to mass.

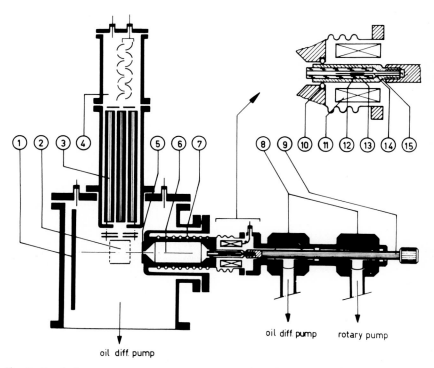

Fig. 1. Pyrolysis-mass spectrometry apparatus: 1, liquid-nitrogen-cooled screen; 2, ion source; 3, quadrupole head; 4, electron multiplier; 5, gold diaphragm; 6, expansion chamber; 7, heating element; 8, three-way metal ball valves; 9, sample probe; 10, Viton O-ring; 11, hf coil; 12, sample region; 13, glass reaction chamber; 14, ferromagnetic pyrolysis wire; 15, reaction tube holder.

The ionization takes place by impact of 15-eV electrons. Mass spectrometers are usually operated at electron energies of 50 to 70 eV to obtain maximum yield of ions. At this electron energy, however, most of the interesting large molecules are not only ionized but also seriously fragmented, thereby complicating the analysis. In spite of the relatively low ion yield obtained with low-voltage electron impact ionization, the resulting mass spectrum is more characteristic because of the presence of large molecular ions and the absence of a bulk of small fragment ions.

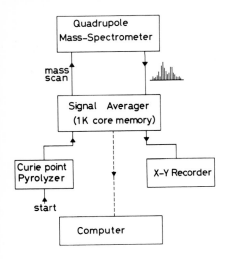

Fig. 2. Block diagram of the experimental setup.

With our Py-MS system mass peaks over m/e 140 were rarely observed for most organic samples analyzed thus far. Although the transmissivity of the quadrupole decreases for larger ions, the adsorption on the wall of the expansion chamber seems to be the limiting factor. To minimize this adsorption, the chamber is kept at 150°C.

Since there are few interesting pyrolysis products to be found below m/e 30, we generally record the mass spectrum between m/e 30 and 140. The mass spectrometer successively scans the relative abundances of the ions, and assuming that the composition of the mixture analyzed is constant in time, one mass scan may give reliable information on this composition. When a mixture is formed during a pyrolysis process and when the products have to diffuse out of the expansion chamber, the composition of the product flow admitted to the mass spectrometer changes considerably in time. Therefore, repeated fast mass scans are performed. The signals are accumulated for each mass during the successive scans; therefore, the final accumulated spectrum represents time-averaged intensities, proportional to the overall composition of the pyrolyzate. The measuring cycle and data recording are regulated by a Fabritek signal averager, which directs the Riber QMM 17 quadrupole mass analyzer to scan successively the masses in the selected range. The signals from the electron multiplier detector are in the form of pulses, each pulse corresponding to the arrival of a single ion. The pulses are amplified, discriminated against noise, shaped properly, counted, and stored in the appropriate channels of the signal averager. This pulse-counting technique allows high scanning speed and more efficient detection than is possible with analog methods.

The final spectrum, accumulated in the memory of the signal averager, can be displayed on an oscilloscope or recorded on strip chart for visual inspection. A paper punch tape readout enables off-line computer processing of the data. The bacterial samples are directly taken from the blood agar plate on which they were cultured and applied to the ferromagnetic wire used in the Curie point pyrolysis. During pyrolysis the wires are heated up to 610°C in about 0.1 sec by a Fischer Labortechnik pyrolyzer unit of 1.5 kW and 1.1 MHz. Total pyrolysis time was 1 sec. In our experimental arrangements a measurable product flow is admitted for 10 sec to the mass spectrometer, when samples of abour 30 μg are pyrolyzed. Although the real measuring time is thus only 10 sec, the total time needed for one analysis is about 5 minutes.

Most time is taken for the insertion of a new sample in the vacuum system and for the data readout procedure. An analysis rate of one sample per minute can certainly be obtained by the use of a multisample inlet system and on-line data transfer to the computer. Developments in this direction are now being made.

RESULTS AND DISCUSSION

The pyrolysis mass spectra for the streptococcal strain classified as Z_3III (MG 216) and its mutant strain Z_3 appear in Fig. 3. Since the mutant differs from the parent strain by the absence of the type III polysaccharide antigen in the cell wall, this difference should be confirmed by the pyrolysis of the purified type III polysaccharide. The analysis of the polysaccharide indicates (see Fig. 3) that this antigen is at least qualitatively responsible for the differences observed in the pyrolysis mass spectra of these bacterial strains. This example, for a well-defined biological system, indicates that the mass spectrum of the pyrolyzate reflects the composition of the original sample.

For these analyses, however, another sampling technique was used. Fine suspensions of freeze-dried samples were made in carbon disulfide by mild sonification. Small drops of these suspensions were brought on the ferromagnetic wires used in Curie point pyrolysis. After evaporation of carbon disulfide, 30 to 40 μg of the sample remained tightly coated on the wire. A detailed description of this technique can be found elsewhere (5).

The mass spectra of the pyrolyzates are considered here as fingerprints, although chemical information can be derived from the spectra (4). In the limited group of bacterial strains analyzed thus far, the reproducibility of the fingerprints, relative to the differences observed in the spectra of nonidentical strains, enables the recognition, at least by eye, of each member of this group. Characteristic patterns are observed, for instance, in the range m/e

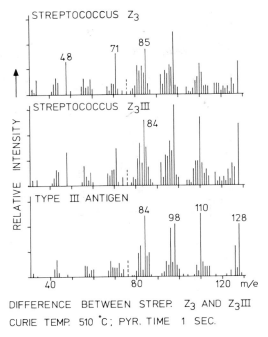

DIFFERENCE BETWEEN STREP. Z_3 AND Z_3III
CURIE TEMP. 510 °C; PYR. TIME 1 SEC.

Fig. 3. Comparison of pyrolysis mass spectra of the streptococcus strain Z_3III and the mutant strain Z_3, lacking the type III polysaccharide.

78–87, as is shown for a number of bacterial strains in Fig. 4. In a representative example of the reproducibility of the spectra (Fig. 5), two independent analyses of *Streptococcus faecalis* (D/D 76) are given. The analyses were performed one week apart, and the samples were taken from different cultures. Most peaks reproduce within 15 to 20%, although larger variations are also observed, especially with peaks of low intensity. As an aid in visualizing the variations in the spectra, for each mass, the ratio of corresponding ion intensities is plotted at the bottom of Fig. 5. In case of perfect reproducibility, including culture growth and sample handling, as well as pyrolysis technique and spectrometric analysis, the ratios must be equal for each pair of peaks and must correspond to the relative sample weights used. In practice, however, the values are scattered around this ideal value, and the mean deviation indicates roughly the degree of reproducibility.

Variations observed in the spectra of different strains are illustrated in Fig. 6 by *Streptococcus mutans* C67-1 and *Streptococcus faecalis* (D/D 76). Distinct

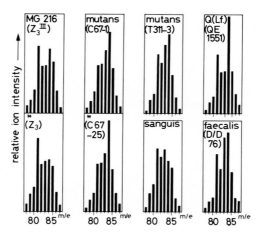

Fig. 4. Characteristic Py-MS patterns observed for different streptococcal strains; asterisk indicates mutant strain.

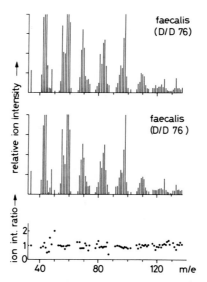

Fig. 5. Comparison of Py-MS spectra of replicate cultures of *Streptococcus faecalis* (D/D 76). Ion intensity ratios were calculated by dividing the ion intensities in the lower spectrum by the corresponding intensities in the upper spectrum.

differences are observed at m/e 50, 69, 110, and 131, whereas the variations at the other masses are in the order of 30%. These findings are visualized by the ion intensity ratios plotted in Fig. 6. To correct for sample weight, the spectra are normalized relative to each other by scaling one of the spectra in such a way that most ion intensity ratios have a value in the range of 0.9–1.1. (see Figs. 5 and 6).

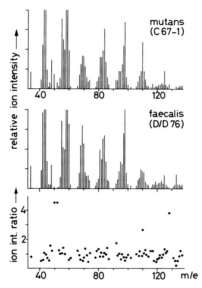

Fig. 6. Comparison of Py-MS spectra of different streptococcal strains. The ion intensities in the upper spectrum were divided by the corresponding intensities in the lower spectrum to obtain the ion intensity ratios.

The visual inspection and comparison of the spectra is rather tedious, especially with regard to the analysis rate of one sample per 5 minutes. Therefore, a procedure was developed for computer matching the spectra of unknown strains to a file of reference spectra. The reference file contains single fingerprints of streptococcal strains analyzed thus far. A numerical value for the correspondence between a spectrum of an unknown strain and a file spectrum is calculated in the following way.

First the spectra are normalized to each other, as outlined previously. The ion intensity ratios are calculated, too, but now the minor peak intensity is always divided by the corresponding major one. This results in ratios smaller than or equal to 1, and the arithmetic mean of the ratios is taken as the degree of correspondence between the spectra compared. In case of perfect matching, the degree of correspondence is 1, whereas marked differences between spectra give rise to a lower correspondence value.

The correspondence value for two spectra of the same strain, however, can be considerably smaller than 1 because of the relatively poor reproducibility of certain peaks. Especially peaks of low intensity are responsible for such variations. Therefore, a selection of peaks is made for the numerical comparison of the spectra. Peaks exhibiting relatively large variations in the spectra of different strains, compared with the variations observed in spectra of replicates, were selected. Based on a selection of 25 peaks, the fingerprints of *S. faecalis* (D/D 76), shown in Fig. 6, have a degree of correspondence of .85, whereas the correspondence between the spectra of *S. mutans* C67-1 and *S. faecalis* (D/D 76) in Fig. 7 is .64.

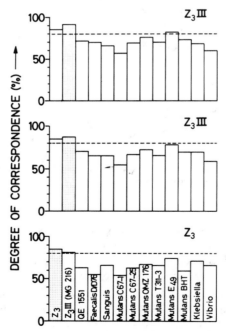

Fig. 7. Bar graphs of computer-calculated degrees of correspondence between spectra of Z_3III and Z_3 and a file of reference spectra. The strains represented by the file spectra are indicated in the lower bar graph.

In our experience, working with a limited set of 25 peaks, spectra of replicate cultures of most strains have a degree of correspondence of .80 or more. A correspondence value of .80 or more observed between any two spectra, therefore, indicates possible identity of the bacterial strains represented by these spectra.

In Fig. 7 the degrees of correspondence between a fingerprint of *Streptococcus* Z_3 and a number of reference spectra are presented in the form of a bar graph. The graph reveals that the best correspondence is obtained between the analyzed Z_3 strain and the reference spectrum of Z_3. On the other hand, an exact identification is not obtained, because the comparison of Z_3 with the reference spectrum of the parent strain Z_3III also yields a correspondence value of more than .80.

The upper part of Fig. 7 gives the results of matching two spectra of Z_3 III against the same file of reference spectra. Two, respectively three, spectra in the file show a reliable correspondence with the analyzed Z_3III strains. However, in both cases the highest degree of correspondence is obtained with the reference spectrum of Z_3III.

Some results obtained by computer matching of unknown strains to a file of reference spectra are illustrated in Figs. 8 and 9. A coded strain, later disclosed to be a *Streptococcus sanguis*, is markedly differentiated from the other streptococci, as Fig. 8 indicates. An example of close relation between a group of strains is given in the lower part of Fig. 8. The mutans C67-25 is just differentiated, but the other mutans strains are quite close.

A *Klebsiella* strain was included in the file to provide an example of differentiation between unrelated species. Also, identifying a coded strain as a *Klebsiella* proved not to be difficult, as can be seen in Fig. 9. In the lower graph all correspondences are less than .80. The sample examined was an obvious contaminant on a culture plate of *Streptococcus faecalis* and indeed could not be identified within our limited file of reference spectra.

Although the matching procedure is based on simple statistical data processing, it performs satisfactorily in a limited group of bacterial strains. The candidate for identification of an unknown strain is sorted out, although if there are more candidates the last step in the identification has to be made by eye. In the analysis of a series of 15 double-blind coded strains, however, 12 strains were identified correctly by the best match, as defined by the highest relative degree of correspondence. Nevertheless, it is clear that more refined pattern recognition procedures must be used when extended reference files are built up.

On the basis of our experience with the identification of bacterial strains by pyrolysis mass spectrometry and the results discussed in this paper, we feel justified in drawing the following conclusions:

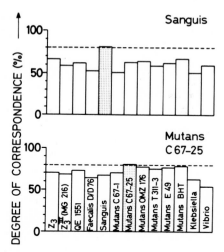

Fig. 8. Bar graph of computer-calculated degrees of correspondence between a sanguis and mutans C67-25 spectrum and a file of reference spectra.

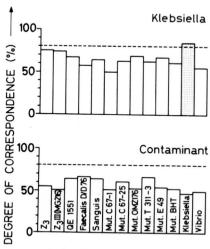

Fig. 9. Bar graph of computer-calculated degrees of correspondence between a *Klebsiella* and contaminant spectrum and a file of reference spectra.

190

1. Fast and reproducible fingerprinting by Py-MS is possible.
2. Analysis rates of one sample per minute may well be feasible.
3. Sample preparation can be reduced to a minimum by taking the sample directly from the culture plate.
4. Generally, less than one colony is needed in the analysis.
5. Data obtained by Py-MS are well suited to computer identification.

Currently we are extending these studies to such other bacterial strains as *Klebsiella* and *Mycobacteria*. Full automation of the system, including on-line computer processing, is envisaged.

Acknowledgments

The authors thank the technical staff at the FOM Institute for expert assistance in these investigations. Furthermore, they are indebted to Professor Dr. J. Kistemaker and Dr. A. J. H. Boerboom for constructive criticism and helpful discussions.

The research was supported by the Organization for Fundamental Research on Matter (FOM) and by the Dutch Ministry of Health

REFERENCES

1. P. G. Simmonds, G. P. Shulmann, and C. H. Stembridge, *J. Chromatogr. Sci.*, **7**, 53 (1967).
2. H. R. Schulten, H. D. Beckey, H. L. C. Meuzelaar, and A. J. H. Boerboom, *Anal. Chem.*, **45**, 190 (1973).
3. H. L. C. Meuzelaar and P. G. Kistemaker, *Anal. Chem.*, **45**, 587 (1973).
4. H. L. C. Meuzelaar, M. A. Posthumus, P. G. Kistemaker, and J. Kistemaker, *Anal. Chem.* **45**, 1546 (1973).
5. H. L. C. Meuzelaar and R. A. in´t Veld, *J. Chromatogr. Sci.*, **10**, 213–216 (1970).

CHAPTER 12

Identification and Grouping of Bacteria by Numerical Analysis of their Electrophoretic Protein Profiles

K. KERSTERS AND J. DE LEY

ABSTRACT

Polyacrylamide gel electrophoresis of cell proteins proved to be a valuable tool in the classification and identification of various microorganisms such as *Mycoplasma* (1–4), *Xanthomonas* (5), plant-pathogenic pseudomonads (6), and several Enterobacteriaceae (7). The method suffers, however, from lack of reproducibility between different electrophoretic runs (8). Moreover, visual comparison of electrophoretic protein profiles and tabulation of relative mobility (E_f) values of bands are subjective and are impractical for large numbers of protein extracts.

We report on a new approach to computer-assisted comparisons of electrophoretic bacterial protein patterns. The method consists of the following major steps:

1. Soluble protein extracts and electrophoretic protein patterns on polyacrylamide gel are obtained under rigorously controlled conditions, to increase reproducibility. For meaningful and reliable computer treatment, it is absolutely imperative that all experiments be carried out with the greatest possible care.

2. Densitometric tracings of these profiles are compared and grouped by computerized calculations. Such comparisons take into account relative mobility, sharpness, and relative protein concentration of bands and zones between bands. The method we propose is only useful when large numbers of strains are to be compared. Small numbers of strains can be compared much more conveniently by visual or photographic means.

EFFECT OF SEVERAL FACTORS ON REPRODUCIBILITY OF PROTEIN PROFILES

We selected two groups of organisms for closer study: *Zymomonas mobilis*, because its protein profile has very sharp bands, and *Agrobacterium*, in which the protein bands are more diffuse. The following factors have little influence on the reproducibility: growth time from log phase up to early stationary phase; the method of cell disrupture (French pressure cell, Ultrason, or MSK-shaker); 90 to 120 minute electrophoresis time; 50 to 150 μg of protein applied per gel cylinder. The cell-free extract can be stored up to 1 month at $-12°$C.

The following factors decrease reproducibility considerably. For the centrifugation of cell-free extracts in an ultracentrifuge at $100,000 \times$ g, protein concentration, both volume of sample and centrifugation time must be kept constant. Protein patterns of the same microorganisms grown in different media can be drastically different. Stringent temperature constancy during polymerization of acrylamide and during electrophoresis, and inclusion of the reference proteins ovalbumin and thyroglobulin improve the reproducibility. Coelectrophoresis of different protein extracts can aid in identification of homologous bands.

Soluble protein was prepared in identical conditions from three different batches each of one *Z. mobilis* and two *A. tumefaciens* strains, and each cell-free extract was analyzed under identical conditions in three separate electrophoretic runs. Clustering levels are excellent at a correlation coefficient $r = .96$ for *Agrobacterium*, but less good at $r = .82$ for *Zymomonas*. Closer inspection of gels and densitometric tracings reveals that small displacements of bands and small differences in sharpness of dense bands occur. Variations are usually more pronounced in the upper third of the gel cylinder and occur in all three soluble protein preparations. Small differences in degree of polymerization of acrylamide may be responsible for these displacements.

It is possible, however, to correct somewhat for these technical variations by a procedure we call the "compensation method." It is a mathematical imitation of the visual compensation, unconsciously performed, when gels are compared with similar band patterns. Numerical treatment of these compensated protein profiles is similar to the ordinary method. Compensation of the *Zymomonas* profiles increases reproducibility to at least $r = .92$. The effect of compensation is less pronounced when the protein bands are less sharp and dense.

EVALUATION, APPLICATION, AND LIMITATIONS

Our method was evaluated with more than 200 genotypically well-known *Agrobacterium* strains: agreement between electrophoretic data on one hand, and DNA : DNA hybridizations and phenotypic clustering on the other, was excellent (Kersters and De Ley, to be published). The classification in the genus *Zymomonas* was likewise solved (Swings and De Ley, to be published).

We found one weakness in our method. When large numbers of protein patterns from a great variety of organisms were compared, we noted exceptional cases of some strains clustering in groups where they did not belong, either phenotypically or genotypically. To eliminate possible mistakes, we make photographic records of the protein profiles of all our strains. After computer clustering, the photographs can rapidly be arranged in the same sequence, and all intruder strains can be picked out visually.

The present contribution is only a preliminary paper. The methods used will be described in detail elsewhere.

REFERENCES

1. S. Razin and S. Rottem, Identification of *Mycoplasma* and other Microorganisms by Polyacrylamide–Gel Electrophoresis of Cell Proteins, *J. Bacteriol.*, **94**, 1807 (1967).

2. S. Rottem and S. Razin, Electrophoretic Patterns of Membrane Proteins of *Mycoplasma*, *J. Bacteriol.*, **94**, 359 (1967).

3. K. A. Forshaw, Electrophoretic Patterns of Strains of *Mycoplasma pulmonis*, *J. Gen. Microbiol.*, **72**, 493 (1972).

4. T. S. Theodore, J. G. Tully, and R. M. Cole, Polyacrylamide–Gel Identification of Bacterial L-Forms and *Mycoplasma* Species of Human Origin, *Appl. Microbiol.*, **21**, 272 (1971).

5. T. A. El-Sharkawy and D. Huisingh, Differentiation among *Xanthomonas* Species by Polyacrylamide–Gel Electrophoresis of Soluble Proteins, *J. Gen. Microbiol.*, **68**, 155 (1971).

6. B. C. Palmer and H. R. Cameron, Comparison of Plant-Pathogenic Pseudomonads by Disc–Gel Electrophoresis, *Phytopathology*, **61**, 984 (1971).

7. T. G. Sacks, H. Haas, and S. Razin, Polyacrylamide–Gel Electrophoresis of Cell Proteins of Enterobacteriaceae, *Israel J. Med. Sci.*, **5**, 49 (1969).

8. J. A. Morris, The Use of Polyacrylamide–Gel Electrophoresis in Taxonomy of *Brucella*, *J. Gen. Microbiol.*, **76**, 231 (1973).

COMPUTER–ASSISTED APPROACHES TO THE IDENTIFICATION OF MICROORGANISMS

Basic Principles in Computer-Assisted Identification of Microorganisms

H. G. GYLLENBERG AND T. K. NIEMELÄ

SUMMARY

The identification logic presented is a combination of the principles of Dybowski and Franklin (1) and Lapage et al. (7). Accordingly, four different "identification states" are possible: (1) *identification* of an unknown as member of a given group, which requires that both absolute and normalized probabilities for that group exceed predefined limits; (2) *intermediate* state as memeber of two or more groups, which is indicated by an absolute probability that exceeds the limit of group membership, but a normalized probability below a predefined limit; (3) *neighborhood* to a given group—the normalized probability may or may not exceed given limit, but absolute probability remains below the limit for group membership, exceeding, however, a secondary neighborhood limit; and (4) the unknown is an *outlier*—the absolute probability does not exceed either membership or neighborhood limits.

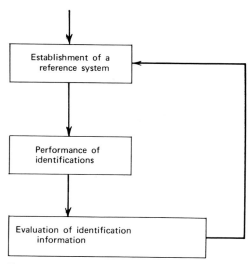

Fig. 1. System flow chart.

words, the reference system is defined by the kind of identification data looked for.

The reference system is most conveniently constructed as a matrix in which each group is defined by the relative frequencies of the different characters. Since this matrix constitutes the main framework for identification, it can be referred to as the *identification matrix*. The identification matrix condenses the relative frequencies of each character for each group that has been obtained as result of the applied classification. Insofar as the characters are dichotomous (i.e., the alternative outcomes are only two; e.g., "+" or "−"), the identification matrix is an $n \times m$ matrix, where n is the number of groups and m the number of characters. In a dichotomous system it is enough to indicate the "+" frequency in the matrix, since the "−" frequency is easily obtained by subtraction. [If the matrix is denoted C, its element c_{ij} defines the frequency of "+" in group i as to character j ($i = 1, \ldots, n, j = 1, \ldots, m$), then the corresponding frequency of "−" is simply given by $1 - c_{ij}$.]

When multistate characters are employed, the structure of the matrix becomes more complex. The only solution is to include the frequencies of all alternative characters in the matrix. This situation has been discussed in some detail by Gyllenberg, Niemelä, and Niemi (5) in connection with the ISP material.

As soon as the identification matrix has been established, it can be continuously utilized for the identification of unknown isolates and cultures

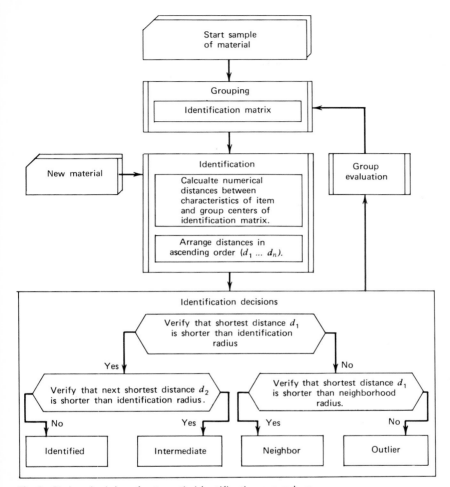

Fig. 2. Basic principles of automatic identification procedure.

(cf. Fig. 2). Different alternative principles are possible for the identification procedure, however, and these are the next subject of detailed discussion.

IDENTIFICATION

General

All data in the identification matrix are treated as equal, no matter what identification principle is applied. This implies that identification can be considered as a means to evaluate the "overall similarities" between the unknown item to be identified and the alternative group centroids given by the identification matrix. (A typical identification matrix is presented in Table 1).

The first main point in our approach to identification is a combination of the identification criteria of Dybowski and Franklin (1) on the one hand, and of Lapage et al. (7) on the other. According to Dybowski and Franklin, a strain or isolate can be identified as belonging to a given group only when the probability of belonging calculated for the group in question exceeds a predefined limit. The principle defined by Lapage et al. presumes that identification can be accepted only when the group in question provides an exclusive alternative. The criterion is that the probability calculated for the group considered must exceed a predefined relative share of the sum of absolute probabilities for all groups in the identification matrix. This relative share is called *normalized probability*.

Our second point is that a combination of both principles just referred to provides *four* different "identification states."

1. The criteria for identification of both Dybowski and Franklin and Lapage et al. are fulfilled. This implies that the probability of belonging to a given group exceeds a predefined absolute value. At the same time, the given group constitutes an exclusive alternative, which means that the normal-ized probability of belonging to this group exceeds a predefined limit for normal-ized probability. If both mentioned conditions are fulfilled, it seems reason-able to consider the item in question (be it a living or nonliving item) to be *identified*.

2. The criterion of Dybowski or Franklin is fulfilled, but that of Lapage et al. is not. In such cases the unknown item has been considered to be an *intermediate* to two or more groups.

3. The criterion of Dybowski and Franklin is not fulfilled, but the probability of belonging exceeds another less rigidly defined border, and the group in question may or may not constitute an exclusive alternative. In this case the unknown to be identified can be defined as a *neighbor* to the group to which it bears the most pronounced resemblance.

4. Finally, the criterion of Dybowski and Franklin is not at hand and the unknown to be identified does not even fall within the "second" border when assessed in terms of absolute probability of belonging. In this case the unknown does not belong to any one of the defined groups and can be considered to be an *outlier*.

The four "identification states" as defined here are illustrated from two different points of view in Fig. 2 and 3.

Table 1 An Example of Identification by Probabilities: Parameters L_{1i}, L_{2i}, and b_i ($i = 1, \ldots, 6$), and Identification Matrix for 6 Groups and 24 Characters

GROUP *****	LIMIT 1 *******	LIMIT 2 *******	BEST POSSIBLE LOG. PROBABILITY ****************
1	3.8274	5.2192	1.5016
2	4.1111	5.5029	1.7852
3	3.2606	4.6524	0.9348
4	3.6830	5.0748	1.3571
5	3.3353	4.7271	1.0095
6	3.3719	4.7637	1.0461

GROUP CENTERS:

TEST	1	2	3	4	5	6
1	0.04	0.00	1.00	1.00	1.00	0.90
2	0.00	0.00	0.98	0.82	0.42	0.90
3	0.00	0.00	1.00	1.00	1.00	1.00
4	0.04	0.24	1.00	0.27	0.17	0.70
5	0.00	0.76	0.00	0.00	0.00	0.00
6	0.74	0.88	0.00	0.00	0.92	0.85
7	0.00	0.59	0.63	0.36	0.92	0.10
8	0.87	0.18	0.00	0.36	1.00	1.00
9	1.00	1.00	1.00	1.00	1.00	1.00
10	0.00	0.00	0.96	0.64	0.75	0.55
11	0.83	0.00	0.00	0.00	0.00	0.00
12	0.04	0.59	0.96	0.00	0.17	0.70
13	0.39	0.35	0.67	1.00	1.00	0.80
14	0.04	0.65	0.96	0.64	1.00	0.85
15	0.09	0.00	0.00	0.00	0.17	0.00
16	0.22	0.00	0.74	0.91	1.00	0.95
17	0.13	1.00	0.89	0.91	1.00	1.00
18	0.57	0.65	0.78	0.91	0.92	0.95
19	0.78	0.94	0.96	1.00	1.00	1.00
20	0.35	1.00	0.96	1.00	1.00	1.00
21	0.04	0.59	0.85	0.18	0.42	0.00
22	0.91	0.94	1.00	1.00	1.00	1.00
23	0.96	0.00	1.00	0.91	1.00	1.00
24	0.83	0.88	0.89	0.91	1.00	0.00
SIZE:	23	17	27	11	12	20

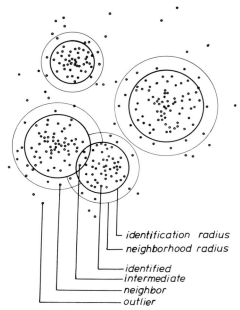

Fig. 3. Illustration of four identification states.

Alternative Identification Procedures

The following alternative identification procedures have been studied in some detail:

1. Identification by the principle of probabilities.
2. Identification based on Euclidean distances.
3. Identification by correlation coefficients.

A fourth method has also been tested—namely, the approach provided by information theory. Since our results from studies of this method are still to be evaluated, the following discussion is restricted to the three first-mentioned principles.

Identification by Probabilities

In identifying by probabilities, we assume that the characters considered are qualitative (i.e., that the outcomes for a character constitute mutually exclusive alternatives), and that there is no weight difference between the alternatives. We also assume, theoretically, that the characters are independent of one another. This identification principle can be easily programmed for general treatment of dichotomous information, but with multistate characters; although the general principle is the same, the program needs to be "tailor-made" for each purpose. The coefficients c_{ij} in the identification matrix must be modified to replace the 1 with 0.99, and the 0 with 0.01. When multistate characters are considered, all frequency figures must be modified to give a sum equal to 1 for the coefficients that describe the probabilities of different alternatives.

When this principle is applied, the frequencies c_{ij} can be interpreted as probabilities for alternatives in question. This implies that the sum of the coefficients for the different alternatives must be 1. The replacement of coefficients attaining the value 0 with 0.01 actually means that highly improbable characters are given a small weight that is higher than zero.

Since the groups in the reference system are considered to be polythetic (which follows, because all characters are given equal weight in identification), information is required that establishes how much the "unknown" to be identified is "allowed" to differ from the mean group characters without implementation of a negative identification result. For this purpose two probability limits—L_{1i} and L_{2i}—are defined for each group i ($i = 1,\ldots,n$).

These limits correspond to the decision situations: identified/neighbor, and neighbor/outlier, which have been described. When these limits are defined, the intragroup variation must be considered. The intragroup variation can be described by b, which is obtained by multiplication of the probabilities of the most frequent alternative outcomes for all considered characters. Thus b_i is the highest possible probability of belonging to group i: b_i is less than 1 and can be expressed as follows:

$$b_i = \prod_{j=1}^{m} \max(c_{ij}, 1 - c_{ij})$$

A high intragroup variation gives a low b, whereas the b for homogeneous groups are high. Since the limits L_{1i} and L_{2i} are defined by $L_{2i} < L_{1i} < b_i$, the subjective parameters 1_1 and 1_2 have been introduced ($1_2 < 1_1$), and L_{1i} and L_{2i} are given by

$$L_{1i} = 1_1^m \cdot b_i \quad \text{and} \quad L_{2i} = 1_2^m \cdot b_i$$

where m is the number of characters.

These parameters 1_1 and 1_2 can be interpreted to show how well the characters of an unknown in average have to correspond to the group characters as a basis for acceptance of identification in or neighborhood to that group. This means that the lower 1-figures are chosen, the higher differences as compared with the matrix are accepted. The 1-parameters have to be less than 1; moreover, we must have $1_2 < 1_1$. Typical values are $1_1 = 0.75–0.85$, and $1_2 = 0.65–0.75$. If $1_1 = 0.85$, an unknown must correspond to a given group at least at an 85% average to be identified as a member of that group.

When information about the unknown is lacking for one or more characters, the limits L_{1i} and L_{2i} have to be changed correspondingly, to maintain relevance of the limit values. If, for instance, information for character k is missing, the new limit L_{1i}^* is obtained from

$$L_{1i}^* = \frac{L_{1i}}{1_1 \cdot \max(c_{ik}, 1 - c_{ik})}$$

In practice, missing information for one character has only little effect and may influence decisions only close to the limits L_{1i} and L_{2i}.

The automated identification decision requires computation of the *absolute probabilities of belonging*, p_i, of the unknown for each of the groups in the identification matrix (i.e., $i = 1, \ldots, n$) from

$$p_i = \prod_{j=1}^{m} f_{ij}$$

where f_{ij} is the probability of character j for group i. In practice, p_i is equal to the total combined probability of all characters of the unknown. For dichotomous characters, we have

$$f_{ij} = \begin{cases} c_{ij}, & \text{when the unknown behaves positively } (+) \text{ for character } j \\ 1 - c_{ij}, & \text{when the unknown behaves negatively } (-) \text{ for character } j. \end{cases}$$

The identification decision also requires evaluation of *how much higher* p_i is, compared with probabilities for alternative groups. For this purpose the *normalized probability*, P_i^*, needs to be computed:

$$P_i^* = \frac{p_i}{\displaystyle\sum_{j=1}^{n} p_j}$$

The normalized probability P_i^* indicates the share of group i of the total sum of probabilities for all groups in question Σp_i. To decide that a given item is identified as a member of group i requires that the absolute probability of belonging to group i be sufficiently high; also, however, the normalized probability must be close to 1 for the given item.

If the limit for the normalized probability is set equal to 0.99 (cf. Ref. 7) the identification criteria are:

1. $P_i^* \geqslant 0.99$
2. $p_i \geqslant L_{1i}$

When condition 2 is fulfilled, but condition 1 is not, the unknown in question is considered to be an intermediate. The normalized probability is not utilized when $p_i < L_{1i}$ (neighbors and outliers). If the most probable group alternative is denoted i, the identification decision is defined by the following conditions (cf. also Figures 2 and 3).

Identified	Intermediate	Neighbor	Outlier
1. $P_i^* \geqslant 0.99$	1. $P_i^* < 0.99$		
		$L_{2i} \leqslant p_i < L_{1i}$	$p_i < L_{2i}$
2. $p_i \geqslant L_{1i}$	2. $p_i \geqslant L_{1i}$		

In practice, the probabilities of items for different groups vary within an extremely wide range (e.g., 10^{-1}–10^{-20}); therefore, the program gives the probabilities as the absolute values of their logarithms (i.e., 1 to 20 according the foregoing example). These logarithmic figures can also be interpreted as distances—a small figure corresponds to a small distance from the group centroid, and vice versa. The program prints out the probability limits L_{1i} and L_{2i} in a corresponding form, as well as the highest possible probabilities b_i. For an unknown to be identified, the program prints out the label, the normalized probability for the best group alternative, and group labels for the three best group alternatives and the corresponding distances (or absolute probabilities). The program also lists the characters in which the item in question differs significantly from the behavior of the corresponding group (i) as defined in the identification matrix. Table 2 illustrates a typical printout.

Identification Based on Euclidean Distances

In our second method, identification is based on Euclidean distances between the individual to be identified and the group centroids. The individuals and the group centroids are considered to be points in an m-

Table 2. An Example of Computer Output of Identification by Probabilities

STRAIN ******	TYPE **********	NORMALIZED PROBABILITY ***********	BEST GROUPS/LOGARITHMIC PROBABILITIES **				UNEXPECTED RESULTS ***************
1	IDENTIFIED	1.0000	1/1.502	2/11.68	6/16.39	4/16.71	
2	IDENTIFIED	1.0000	1/3.301	2/14.89	6/18.42	4/18.71	
3	OUTLIER	0.9999	5/5.038	3/9.461	4/9.483	6/9.634	5
4	NEIGHBOR	1.0000	6/4.135	4/10.07	1/10.63	5/11.36	
5	IDENTIFIED	1.0000	1/2.178	2/12.55	6/14.39	4/17.71	
6	OUTLIER	0.8667	1/11.00	2/11.84	4/13.13	5/15.09	9 12 21
7	IDENTIFIED	1.0000	1/1.502	2/11.68	6/16.39	4/16.71	
8	IDENTIFIED	0.9998	3/0.935	4/4.675	5/5.737	6/8.741	
9	NEIGHBOR	0.9997	6/3.410	5/6.887	4/8.587	3/10.23	20
10	IDENTIFIED	1.0000	1/2.292	2/12.81	6/15.67	4/18.71	

11	IDENTIFIED	1.0000	2/2.560	1/11.66	5/12.31	4/13.13	
12	NEIGHBOR	0.9935	5/4.695	3/6.922	6/7.987	4/8.909	5
13	NEIGHBOR	1.0000	1/5.180	2/15.33	4/19.71	6/21.44	12
14	IDENTIFIED	1.0000	6/1.133	5/6.068	4/7.260	3/8.234	14
15	NEIGHBOR	0.8247	5/3.628	4/4.325	6/5.572	1/11.35	
16	IDENTIFIED	1.0000	1/1.502	2/11.68	6/16.39	4/16.71	
17	IDENTIFIED	1.0000	2/2.312	1/12.23	4/18.37	3/18.45	
18	OUTLIER	1.0000	6/5.763	1/11.15	3/11.31	4/11.50	17
19	IDENTIFIED	0.9998	3/0.935	4/4.675	5/5.737	6/8.741	
20	IDENTIFIED	0.9998	3/0.935	4/4.675	5/5.737	6/8.741	
21	IDENTIFIED	0.9939	6/3.209	4/5.492	5/6.264	3/9.649	
22	NEIGHBOR	0.9998	1/4.023	2/7.949	5/8.515	6/8.522	
23	IDENTIFIED	0.9998	3/0.935	4/4.675	5/5.737	6/8.741	
24	OUTLIER	0.9999	3/6.217	4/10.26	6/10.66	1/16.53	14
25	OUTLIER	0.9994	2/6.673	1/9.918	6/16.26	4/18.39	17

dimensional space, and it follows that the characteristics should be quantitative (in special cases; dichotomous qualitative characters can be accepted). In this case the identification matrix contains the means of the characters group by group, and occasionally occurring figures 0 and 1 do not need to be modified as in the preceding method. To get an equal weighting of all characters, their outcomes have to be modified to fall within the same numerical range (e.g., the range $[0,1]$). For dichotomous characters, 0 corresponds to a negative outcome ("$-$"), and 1 to a positive outcome ("$+$"). The group centroids are based on the relative frequencies of positive outcomes.

The intragroup deviation, which has to be considered for identification decisions, is σ_i. For dichotomous characters the deviations can be calculated directly from the identification matrix by application of the formula for the variance of the binomial distribution. In other cases the deviations must be calculated in connection with grouping and must be defined for the identification program. For each group, two radii are defined as multiples of the deviation. The smaller radius corresponds to the decision identified/neighbor, whereas the outer radius constitutes a limit for the decision neighbor/outlier. These limits are given to the program by the parameters 1_1 and 1_2, and the corresponding radii are $1_1 \cdot \sigma_i$ and $1_2 \cdot \sigma_i$.

It follows from this definition of the identification radius that if the distance between two group centroids i and j is

$$d_{ij} < 1_1 \cdot (\sigma_i + \sigma_j)$$

the spheres representing both groups (i and j) are partly overlapping. An unknown individual lying within the intersection of both groups thus constitutes an intermediate.

To facilitate comparison of different alternatives, the distances d_i between the individual items and the group centroid are divided by the group deviation. This gives a *normalized distance*: $D_i = d_i / \sigma_i$. Denoting the shortest distance D_1, and the next shortest D_2, the identification conditions can be described as follows.

Identified	Intermediate	Neighbor	Outlier
1. $D_1 \leqslant 1_1$			
	$D_2 \leqslant 1_1$	$1_1 < D_1 \leqslant 1_2$	$D_1 > 1_2$
2. $D_2 > 1_1$			

The parameters d_i and σ_i are defined by

$$d_i = \sqrt{\sum_{j=1}^{m} \left(c_{ij} - x_j \right)^2} \qquad \text{and} \qquad \sigma_i = \sqrt{\sum_{j=1}^{m} c_{ij} \left(1 - c_{ij} \right)}$$

where m is the number of characters, x_j is the value of the unknown for character j, and c_{ij} is the mean value for character j in group i. The parameters 1_1 and 1_2 are chosen subjectively, as described in the preceding section, considering that $1_1 < 1_2$.

Identification by Correlation Coefficients

The method of identification by correlation coefficients applies primarily to quantitative characters, but dichotomous characters can also be used. The similarity between the individual to be identified and the group is evaluated by the correlation coefficient of the corresponding characters. The significance of the correlation coefficient is tested by the t-test. The superiority of the best group alternative can be examined by statistical testing of significant similarity of the two highest correlation coefficients at the selected level of confidence. The program gives the three best group alternatives and the corresponding correlation coefficients, as well as the confidence levels in percentages. Missing information for an individual in a given character can be replaced by the mean for this character in the whole material. This introduces an error, but a change in the degrees of freedom is avoided.

When desired, the *normalization* of the variables (i.e., the characters) can be performed by transforming the variance mean to zero and the variance to one. In a dichotomous material the variance estimates needed for normalization can be obtained directly from the identification matrix by application of the formula for calculation of the binomial distribution. In other cases the variances must be calculated when grouping is performed, and the obtained figures must be given to the identification program. The normalization, if performed, has to be carried out both for the identification matrix and for the unknown individuals to be identified.

For identification, the correlation coefficients between the individual in question and the alternative groups are calculated: $r_i(i = 1, \ldots, n)$. The two highest coefficients r_I and r_{II} are considered, and the figures t_I and z, corresponding to them, are calculated. From tables included in the program, the corresponding significance figures are then defined. On this basis, we either accept or reject the zero-hypotheses, namely:

1. $H_0 : r_I = 0$ (t-test)
2. $H_0 : r_I = r_{II}$ (double-sided test)

For evaluation of the first hypothesis, two limits (T_1 and T_2) are predefined. These limits correspond to the identification decisions identified/neighbor and neighbor/outlier. For the second hypothesis, the limit Z is given. The conditions for the identification decisions are the following:

Identified	Intermediate	Neighbor	Outlier
1. $t_I \geqslant T_1$	1. $t_I \geqslant T_1$		
		$T_2 \leqslant t_I < T_1$	$t_I < T_2$
2. $z \geqslant Z$	2. $z < Z$		

The correlation coefficient between the individual x and the group centroid c_i is given by

$$r_i = \frac{\sum\limits_{j=1}^{m} x_j c_{ij} - \left(\sum\limits_{j=1}^{m} c_{ij}/m \right) \sum\limits_{j=1}^{m} x_j}{\left\{ \left[\sum\limits_{j=1}^{m} x_j^2 - \left(\sum\limits_{j=1}^{m} x_j \right)^2 /m \right] \left[\sum\limits_{j=1}^{m} c_{ij}^2 - \left(\sum\limits_{j=1}^{m} c_{ij} \right)^2 /m \right] \right\}^{1/2}}$$

The t-figure corresponding to r_i is

$$t_i = \frac{r_i \sqrt{m-2}}{\sqrt{1 - r_i^2}}$$

The z-figure corresponding to r_i is

$$z_i = \frac{1}{2} \ln \frac{1 + r_i}{1 - r_i}$$

The z-figure indicating the difference between two correlation coefficients is

$$z = \frac{z_1 - z_2}{\sqrt{2/(m-3)}}$$

where

m = number of characters

c_{ij} = mean of character j in group i

x_j = value of individual x in character j

GROUPING AND FEEDBACK

As pointed out earlier different alternatives are open for creation of a relevant identification matrix. In our study of the ISP material (5), fundamental identification logics were applied to classification. Although our treatment of the ISP material will be published elsewhere, a brief discussion of "classification by identification" may be worthwhile in this connection.

Particularly when Euclidean distances are applied, the individual cultures can be imagined as points in a multidimensional space. If the occurrence of well-definable groups is supposed, we can conclude that some regions of the given space must be more densely occupied than others, and the aim of any classification should be to find just the dense clusters of points in the space.

It follows from these considerations that classifications should logically start from the densest regions of the character space. In our earlier report (5), a special index was used to indicate how the individual strains were located in the character space in relation to each other. This index (Q) was given by $Q_i = 2 \cdot I + N$, where I is the number of items that are completely identical with item i and N is the number of items that differ in only one character.

A high Q-index shows that the item in question is located in a densely occupied region of the character space, whereas a low Q-index points to location in a sparsely occupied region.

Grouping or classification can then be initiated by imagining that the item possessing the highest Q-index is representative of a group centroid. An L_1-limit can thus be defined, and this allows identification against a reference set composed of only one, still-imaginary group. Since the selected group centroid by definition lies in a densely occupied region of the character space, several items are likely to fulfill identification conditions, also when L_1 is defined very rigidly. As soon as the first group in the system increases, it becomes possible to compute representative frequencies for the outcomes of different characters.

Simultaneously with the collection of the first group, the method investigates possibilities for the creation of further groups. This implies utilization of the parameter L_2 to avoid having the new groups too close to the first one or to each other. With this restriction in mind, the basic material (cf. Fig. 2) can be treated successively until no further groups are produced.

This method of classification by identification proceeds through several iteration cycles, each cycle starting from the situation produced by the previous cycle. In this way a grouping is created successively with adjustment and correction during each cycle. The grouping approaches stability, which means that the number of changes (i.e., introduction of new groups and rearrangement of already placed items) decreases for each cycle. When no further changes occur, classification is completed. In the ISP study five to seven iterations were usually needed to reach practically sufficient stability.

As indicated by Fig. 1, information from identification should be continuously utilized for an adjustment and correction of grouping, thus also the identification matrix. Rearrangements may concern both fusion and fission of preexisting groups. However, the parameters for automated steering of this rearrangement process are not easy to define. The problem is now being studied in our group, and we shall not discuss it further here.

APPLICATIONS

As we mentioned, the identification methods described have been tested on two distinctly different materials: (a) 448 published culture descriptions from the ISP study, and (b) 110 + 223 isolates from clinical specimens. Since the results obtained with these materials will be discussed in detail elsewhere, we make only a a few comments in the present context.

The main question concerning application is of course the determination of the extent to which data from automated identification agree with identification information of other kinds. Our reference matrix for the ISP material included 15 groups. Some were found to lie close together (short distances between group centroids). This was the case for our groups 3, 6, and 8. Ettlinger, Corbaz, and Hütter have listed a number of cultures that on the basis of their scheme belong to the *Streptomyces griseus* group (2). The following 10 organisms of the material of Ettlinger et al. were also included in the ISP study (identifications according to our system are indicated in parentheses):

Streptomyces albidus	(identified: group 3)
S. albus	(identified: group 8)
S. alni	(neighbor: group 3)
S. citreus	(neighbor: group 8)
S. coelicolor	(neighbor: group 6)

S. californicus	(identified: group 6)
S. filipiensis	(outlier, closest to group 2)
S. fradiae	(intermediate: groups 14, 11)
S. olivochromogenes	(neighbor: group 15)
S. praecox	(intermediate: groups 3, 4).

Since 70% of the cultures are clustered around groups 3, 6, and 8, a fairly good agreement with the conclusions of Ettlinger et al. was found.

Recent data of Okanishi, Akagawa, and Umezawa provide a basis for a similar comparison (g). Okanishi et al. have grouped part of the ISP material on the basis of DNA homology with *Streptomyces griseus* and *Actinomyces globisporus*. Among the groups defined by Okanishi et al., "subgroup 5" is large enough to provide a basis for comparison:

Streptomyces badius	(identified: group 8)
S. griseinus	(identified: group 3)
S. fluorescens	(identified: group 3)
S. californicus	(identified: group 6)
S. vinaceus	(outlier, closest to group 14)
S. cavourensis	(outlier, closest to group 6)
S. albidus	(identified: group 3)
S. griseobrunneus	(identified: group 6)
S. bikiniensis	(neighbor: group 8)
S. rubiginosohelvolus	(identified: group 3)
S. alboviridis	(identified: group 3)
S. microflavus	(identified: group 3)

In this case not less than 6 out of 12 cultures were identified as belonging to group 3; furthermore 2 cultures were identified as belonging to group 6, one culture was identified with group 8, and one was found to be a neighbor to group 8. Of the two outliers, one actually showed affinity to group 6. Accordingly, an agreement between our results and those of Okanishi et al. seems to be obvious.

For the clinical material, we simply want to present a comparison of the outcomes of the three identification methods when applied to the same isolate material. As can be seen from Table 3, there is a distinct agreement between the decisions produced by identification on the basis of probabilities and identification utilizing Euclidean distances. Identification by correlation coefficients in this material eliminated the neighbors almost completely but produced a high number of intermediates. The highly similar distributions of data with the different methods, which are independent of one another, can be considered to be evidence of the validity of our approach to identification.

Table 3 Comparison of Different Identification Principles Applied to Clinical Isolate Material

Group[a]	Identified	Neighbor	Intermediate	Outlier	Σ
Probability					
1	44	12		8	64
2	20	7		5	32
3	35	6	2	6	49
4	2	4	2	3	11
5	12	7	6	4	29
6	28	5		5	38
Σ	141	41	10	31	223
	63.2%	18.4%	4.5%	13.9%	
Correlation					
1	46		13	2	61
2	26	1	5	2	34
3	34		18		52
4	1		9		10
5	3		23	1	27
6	27		12		39
Σ	137	1	80	5	223
	61.5%	0.5%	35.8%	2.2%	
Correlation (normalized)					
1	48	2	9	5	64
2	30	1	1	5	37
3	35		15		50
4	1		8		9
5	4	2	15	2	23
6	33		6	1	40
Σ	151	5	54	13	223
	67.8%	2.2%	24.2%	5.8%	

Table 3 Comparison of Different Identification Principles Applied to Clinical Isolate Material

Group[a]	Identified	Neighbor	Intermediate	Outlier	Σ
Euclidean Distance					
1	45	14		5	64
2	24	7		4	35
3	33	2	4	9	48
4	6	7	1	6	20
5	8	2	5	4	19
6	26	8		3	37
Σ	143	39	10	31	223
	64.0%	17.6%	4.5%	13.9%	

[a] Groups: 1, *Pseudomonas*; 2, *Proteus*; 3, *E. coli*; 4, *E. coli* and *Enterobacter*; 5, *Enterobacter*; 6, *Klebsiella*.

REFERENCES

1. W. Dybowski and D. A. Franklin, Conditional Probability and the Identification of Bacteria. A Pilot Study, *J. Gen. Microbiol.*, **54**, 215–229 (1968).

2. L. Ettlinger, R. Corbaz, and R. Hütter, Zur Systematik der Actinomyceten. 4. Eine Arteinteilung der Galtung *Streptomyces* Waksman et Henrici, *Arch. Microbiol.*, **31**, 326–358 (1958).

3. H. G. Gyllenberg, A Model for Computer Identification of Microorganisms, *J. Gen. Microbiol.*, **39**, 401–405 (1965).

4. H. G. Gyllenberg, Numerical Methods in Automatic Identification of Microorganisms, *Bull. Ecol. Res. Commun. (Stockholm)*, **17** 127–133 (1973).

5. H. G. Gyllenberg, T. K. Niemelä, and J. S. Niemi, A Model for Automatic Identification of Streptomycetes, to be published.

6. H. G. Gyllenberg, T. K. Niemelä, J. S. Niemi, Hannele Jousimies, and Annele Hatakka, manuscript in preparation, 1973.

7. S. P. Lapage, S. Bascomb, W. R. Willcox, and M. A. Curtis, Computer Identification of Bacteria, in *Automation, Mechanization and Data Handling in Microbiology*, A. Baillie and R. J. Gilbert Eds., Academic Press, London, 1970, pp. 1–22.

8. S. I. Niemelä and H. G. Gyllenberg, Application of Numerical Methods to the Identification of Microorganisms, *Spis. Prirodoved. Fak. Univ. J. E. Purkyne v Brne, Ser. K 43*, **495**, 279–289 (1968).

9. M. Okanishi, H. Akagawa, and H. Umezawa, An Evaluation of Taxonomic Criteria in Streptomycetes on the Basis of Deoxyribonucleic Acid Homology, *J. Gen. Microbiol.*, **72**, 49–58 (1972).

10. E. W. Rypka and R. Babb, Automatic Construction and Use of an Identification Scheme, *Med. Res. Eng.*, **9**, 9–19 (1970).

CHAPTER 14

Problems and Approaches in Computer-Aided Analysis of Virus-Infected Cells

HANS. M. AUS, VOLKER TER MEULEN, PETER H. BARTELS, GEORGE L. WIED

SUMMARY

Pattern recognition has been applied successfully to the problem of recognizing cell changes due to malignant cell growth. Application of these techniques to virological studies was feasible because viruses alter their host, its metabolism, and its morphology. Virus-infected cells were measured in a Universal-Mikrospektrophotometer (UMSP-I, Zeiss) with the fast scanning stage in the UV and visible light spectra, the recorded data were analyzed off-line on a large-scale computer. Cell changes in the sequence of a virus infection were investigated in HeLa cells exposed to poliomyelitis virus. Features extracted from the measured data could be used to classify the infected cells. Furthermore, the kinetic parameters associated with the cell infection could be statistically analyzed. This approach to the recognition of cell changes can be applied in virology as a new method of analyzing molecular structure changes and cell alterations resulting from virus infections.

INTRODUCTION

Computer-aided analysis of intracellular patterns of image scan data has been previously reported (1–4). The successful application of this technique is possible because the recorded absorption data vary directly according to the structural changes in the gray values of the cells. Computer algorithms were developed to evaluate these complex, often subvisual, morphological relationships (4). Most of the studies were conducted on tumor cells, to detect malignant cell changes (3). Cell changes are not limited to structural differences as found in malignancies; however, they also occur in viral-infected cell cultures. Viral infection is accompanied by biological changes in the cytoplasm and/or nucleus of the host. The process of viral multiplication leads to profound alterations of the host metabolism which can appear as structural or textural differences in the scanned data.

Our first study (10) using a carrier culture infected with Moloney leukemia virus indicated that viral infection may cause both quantitative and qualitative changes of magnitude useful in computer-applied analysis. The objective of this study was to analyze the sequence of infection in a cell culture infected with poliomyelitis virus in an attempt to statistically describe the kinetics of virus multiplication.

MATERIAL AND METHOD FOR AUTOMATED CYTOPHOTOMETRY

Virus:

Type 1 polio virus, strain Mahoney, propagated in HeLa cells, was used in this study.

Cell Cultures:

HeLa cells were grown in Minimum Essential Medium supplemented by fetal calf serum. The cells showed no sign of contamination with pleuropneumonialike organisms (PPLO).

Cell Preparation for Cytophotometry:

Round quartz coverslips with a diameter of 17 mm (Carl Zeiss, Oberköchen, Germany) were placed in petri dishes. To avoid a confluent cell sheet, a low cell number per coverslip was used, single cells being mandatory for cytophotometric measurements in the Universal-Mikrospektrophotometer (UMSP-I, Zeiss). Each coverslip was seeded with 0.2 ml of HeLa suspension containing 1×10^5 cells/ml. The petri dishes were subsequently incubated at 37°C in a CO_2 atmosphere, after 5 hours, 5 ml of growth medium was added to each petri dish. The coverslips were washed twice in $1 \times$ PBS 24 hours later, and the cells were infected with polio type 1 virus at an input multiplicity between 40 and 50. Over a period of 7 hours, at 60-minute intervals, infected and noninfected cells were harvested on the coverslips. One set of coverslips was stained with methylene blue according to the method of Deitch (5); the other set was fixed for 48 hours in 4% buffered formulin (veronal acetate buffer, pH 7.2). The stained or formulin-fixed coverslips were mounted on quartz slides.

Cytophotometry:

A recording microspectrophotometer (UMSP-I) with a fast scanning stage interfaced to a minicomputer (PDP-12, Digital Equipment Corporation, Maynard, Massachusetts.) was used in this study (11). The absorption measurements were carried out in the ultraviolet light at 260 nm and the methylene blue preparations at 610 nm. Thirty cells were scanned for each hour of infection. The recorded cell images were transferred, reformatted, and processed on a UNIVAC 1108 and a CDC 6400.

Data Analysis Programs:

A population of cells deriving from the same source may exhibit different microscopic images that hinder the observation of the common significant class properties. Conversely, insignificant visual similarities may result in the incorrect classification of two cells as identical. However, when measured on a scanning microscope and converted to arrays of digitized gray values, each point of every cell can be analyzed by exactly the same computer algorithms. Complex relationships between the recorded gray values can be precisely evaluated. Furthermore, insignificant visual information can be ignored.

The main sections of the taxonomic intracellular analytic system (TICAS) program applied here (1) appear in Fig. 1. The feature extraction section is described under Results. In this study, the output of the feature extraction section consists of two sets of features with 18 components each, so that each cell is represented by a 36-dimensional feature vector (Fig. 2).

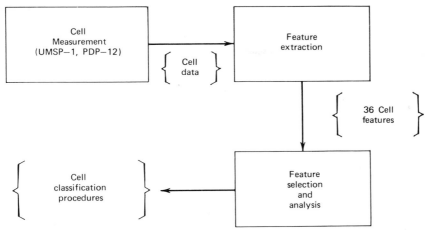

Fig. 1. Block diagram of cell measurement and analysis system.

The feature selection and analysis program section, DISTILL, is a supervised learning program. It examines two sets of feature vectors for components whose properties permit discrimination between two sets of cells that are known to have been derived from two different sets of cells. Here, the two sets of cells are drawn from infected and uninfected cultures at various hours of infection. The program is called a supervised learning program

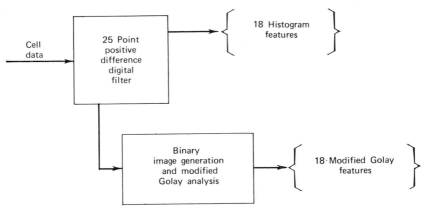

Fig. 2. Block diagram of feature extraction system.

because it learns by examination of the properties of the members of each set of data (also called the training sets) to derive a classification rule, thus to distinguish between members of such sets.

The program section employs two subroutines, DSELECT and EX-TRACT (Fig. 3). DSELECT, rewritten for the UNIVAC 1108 and slightly modified from the version described in the literature to check for continuity of the data, searches for the best discriminating image properties. It is a nonparametric procedure; that is, it not depend on information about the statistical distribution of the data it processes. The procedure tends to select image properties of low correlation. DSELECT may either classify cell images directly or pass the selected image properties to the second subprogram. EXTRACT forms linear combinations of the image properties chosen by DSELECT. It assigns weights that are determined by the means and respective variances for a given image property in the two training sets. For the two sets of features (Fig. 2), described below, two linear combinations result—LINCOMB 1 and LINCOMB 2. These are then processed in a second cycle (Fig. 3) by the same subprogram EXTRACT, resulting in the final composite discriminator, LINCOMB 3.

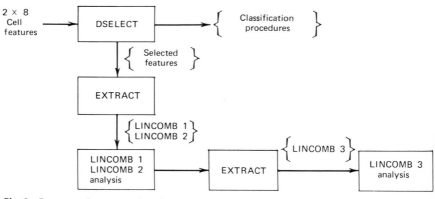

Fig. 3. Feature selection and analysis system.

For each cell the likelihood for being a member of the set of uninfected cells or the set of infected cells is computed. For the entire cell sample the average likelihood for correct classification, $\overline{LR}$, is computed.

In setting up the classification algorithm, the supervised learning section of the master program employed for this study considered every cell of the infected cell sample a priori as being different from the cells of the infected sample.

COMPUTERIZED RESULTS

Infected HeLa cells in the UV microscope exhibit, in the sequence of infection, morphological changes in the cytoplasm. After 2 hours of infection, UV-light-absorbing granules appear in the cytoplasm of some cells (Fig. 4b) which are not present in the uninfected cells (Fig. 4a). In addition cytoplasmic vacuoles develop (Fig. 4b). This phenomenon increases in intensity and is present in every cell at 4 hours after infection (Fig. 5). At a later stage (5–6 hours) these granules disappear, and the cell shows significant structural disintegration. The cytoplasm granularity is less readily apparent in the methylene blue stained preparations.

Figures 6a, 7a, and 8a, which are typical printouts of the measured cells, demonstrate that unequally infected cells (uninfected vs. 4 hours infected, Figs. 6a and 7a) can appear to be visually similar, whereas equally infected cells can appear to be dissimilar (Figs. 7a and 8a).

The basic problem in analyzing the texture associated with the granularity of an infected cell is conceptually represented in Fig. 9a. Gray values reflecting granularity changes due to the infection are masked by the gray values of the structure of the host. Therefore, the first step in the feature extraction program must be to obtain textural gray values that are dependent on viral infection and independent of cell structure. The desired results are given in Fig. 9b. Two-dimensional filtering techniques are one method of calculating local textural differences (9). A 25-point equal-weighted digital filter was used to analyze the HP and PM series (HP is ultraviolet measured cells, PM is methylene blue stained preparations).

The 25-point filter is applied at each point in the cell data where the gray value is greater than 19 (the maximum gray value being 180). The results of subtracting the filter output from the measured gray value (Figs. 6a, 7a, 8a) at each point are presented in Figs. 6b, 7b, and 8b, where only the positive differences have been retained.

One set of extracted features is simple histograms of this positive difference filter (PDF) as in Fig. 2. The second set of features is derived using a modified version of the binary image techniques described in References 6 and 7. Fig. 10 indicates, these techniques tell whether a given point is a member of an edge of a string or is an internal point (8). The binary image of the PDF is obtained by setting all points where the output of the PDF is greater than 1 to a binary 1, and to 0 otherwise. The results of this step is shown in Fig. 6c. The 18 binary image features are described in the Appendix.

Program DISTILL (Fig. 3) is able to distinguish between cells of two different cultures using the extracted features described previously. A summary of the results follows.

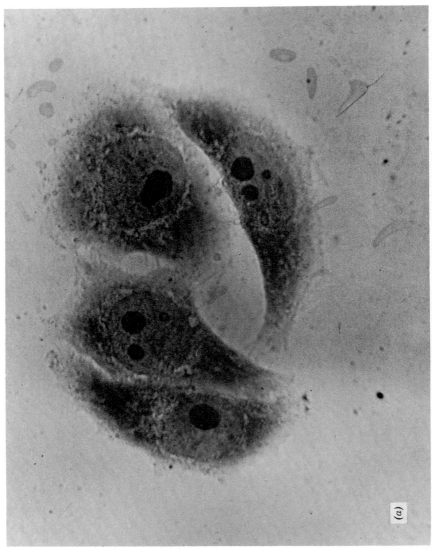

Fig. 4a. Uninfected HeLa cells viewed in the ultraviolet microscope at 260 nm.

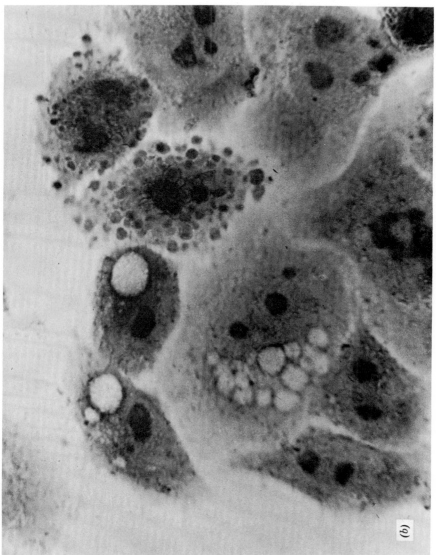

(b)

Fig. 4b. Polio-infected HeLa cells 2 hours after infection, viewed in the ultraviolet microscope at 260 nm.

235

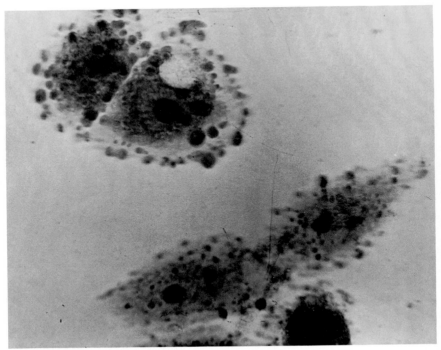

Fig. 5. Polio-infected He*ia* cells 4 hours after infection, viewed in the ultraviolet microscope at 260 nm.

The statistics associated with a histogram and a binary image extracted from the PM series are given in Tables 1 and 2. These two feature components were selected by DSELECT as exhibiting significant differences possibly suitable for a separation of the data sets.

The result of applying the DSELECT–EXTRACT sequence to the two feature vectors extracted from cell cultures of 2, 3, and 4 hours of infection is shown in Table 3. The relative weights were assigned by EXTRACT to the features selected by DSELECT. DSELECT's ability to properly distinguish between the two sets of cells is summarized in Table 4. Using either histogram or binary features, DSELECT is nearly always able to classify properly more than 70% of the PM cells with six or fewer features. The one exception is control versus 0.5 hour of infection, where 50 and 44% of the cells remain unclassified using the histogram and binary image features, respectively. For both measurement series, DSELECT was able to distinguish more cells by using the histogram features than by using the binary image features. The average likelihood (with no optimization) of correct

```
                        4 13  8  7
                     8 19 30 19 20 14
                 10 26 24 30 28 26 12
              3 13 36 31 31 33 35 21
                16 47 32 32 32 36 34  9
                18 40 43 44 25 27 35 18  5
              3 19 35 57 52 33 41 41 15  9
             10 25 39 50 43 47 52 46 34 21
           6 17 30 45 51 47 37 54 42 20 20 10
        3  9 25 33 43 52 47 31 48 49 48 34 17  8
        9 20 33 36 41 55 51 31 66 74 70 62 48 20 12  5
        9 27 3C 40 44 49 37 67 74 76 74 70 58 45 21 18 16 12  6
       10 21 34 42 34 39 49 68 69 75 80 72 64 51 31 21 17 14 14
       10 20 28 41 31 42 62 74 75 78 81 77 73 59 50 21 21 18 22  4
        9 19 28 36 33 49 63 73 80 77 83 82 74 67 56 34 30 21 30  6
   5    8 16 28 29 26 43 67 72 81 80 78 76 75 69 64 29 28 25 30  4
        8 19 30 31 22 37 66 73 82 80 79 79 76 72 64 33 23 26 32  6
        4 18 20 12 21 38 66 75 82 79 82 78 77 76 56 46 39 36 31 15
       10 20 12 26 51 64 74 81 80 77 77 73 72 47 40 43 42 33 25
        7 23 29 30 45 58 71 78 79 78 76 75 59 31 25 44 40 37 28
        6 12 20 24 30 41 66 70 72 73 72 68 35 17 39 50 36 30 32
        5 12 19 21 23 20 48 65 68 69 67 74 62 35 48 43 45 41 35
        4 10  9  5 20 32 26 44 63 60 66 75 75 52 48 32 44 33 29
        5 12 11 26 29 22 24 38 44 58 66 53 45 57 33 43 22 15
              9 15 25 38 51 49 37 43 48 25 38 46 39 42 19 17
              5  7 24 41 51 58 46 25 43 42 47 29 26 38 28 23
             10 20 30 26 52 51 21 41 37 41 37 42 34 23 15
              6 12 21 19 45 47 22 48 45 45 38 38 29 19  7
                 9 16 28 17 30 36 42 30 38 35 31 23 15
                 6  9 14 15 12 34 36 28 30 26 26 18 10
              3  C  5 12  6  4 26 34 28 32 26 21 15  6
              3 18    10    9 12 19 20  6 22  7 11  8  3
                 4     4  8 12 10 11  3 21  7  8  0
                       5  6  5  4  7  6  6  4  3
                                (a)
```

Fig. 6. Computer printout. (*a*) cell data on uninfected HeLa cell scanned by the UMSP-I.

recognition is also given in Table 4. Individual likelihoods for both test series ranged from $1.1:1$ to $10^8:1$ with no optimization.

The resulting linear combinations of the sequence DSELECT–EXTRACT, LINCOMB 1, and LINCOMB 2, are also random variables and can be further analyzed, for example, by EXTRACT. The final linear combination, in this case LINCOMB 3, of the DSELECT–EXTRACT, EXTRACT sequence appears in Table 5. The weights assigned by EXTRACT are a measure of the relative importance of the extracted features in classifying the cells. One example of the statistical properties of the final linear combination (Table 6) indicates clearly that the cells from the control culture and from the 4-hour infected culture, PM series, are distinguishable.

Similar results were obtained using the UV light. An abbreviated summary is presented in Tables 7 to 9. Histogram features played a slightly more important role than the binary image features. Likelihood ratios in the

```
                                                  6  10
                                               3 10 10  8  8  6
                                          4  3  8 24 23 14 16 15  4  3
                                          9 15 15 26 30 25 22 19 15  6
                        11  6             18 19 15 27 35 35 26 30 25 12
                         7 23 19  9  4  7 23 24 19 32 37 33 40 42 29 12
                   3 12 19 19 17 10 16 19 19 30 48 37 23 34 41 27 16  7
          7  7  9  8  9 10 19 13 19 20 24 40 45 42 42 40 41 32 26 12
       3 18 21 22 13 15 19 17 16 14 29 30 36 36 31 45 42 25 18 28 13
         22 33 38 12 11 14 25 18 18 29 38 29 39 51 54 54 28 27 28 15
         19 39 47 28 21 20 19 26 33 24 32 52 59 65 65 58 35 19 23 12
          7 35 45 34 21 21 17 21 17 22 50 67 73 66 72 70 41 16 16 10
          0 27 35 20 11 16 22 28 32 56 68 66 78 76 76 73 61 23 13 11
          0 18 26 17 14  5 10 12 42 60 66 74 81 78 80 74 68 53 25  7
          4 17 19 21 17  5  9 25 59 64 66 77 82 79 81 76 61 67 26  7
         10 18 13 17 21 14  8 21 63 66 72 77 82 79 83 71 59 62 39 11
       5 12 13  4  8 23 18 10 28 65 73 70 75 80 77 80 64 62 58 45 12
     5 20 14 19 13  9 14 19 15 25 60 66 74 76 78 75 79 69 66 56 48 19
  4 18 18 13 14 20 15 17 29 31 29 63 72 69 66 70 67 75 79 69 64 44 22
 15 25 18 22 13 20 19 14 24 27 18 61 71 77 77 77 71 74 80 70 63 29 19  3
3 13 18 29 16 16 23 22 19 19 20  8 42 73 80 76 80 71 78 80 68 55 14 13  4
  8 17 29 25 18 17 20 26 25 13 10 41 69 76 78 73 63 67 71 65 33 28 14
  7 27 34 30  9 13 20 24 25 17 21 41 60 66 74 69 68 67 70 47 34 19
 17 37 31 23  8 13 16 16 25 31 28 36 41 45 48 55 59 66 62 32 26 12
 14 27 18 15 13 15 18 21 33 34 12 24 30 33 37 58 39 28 38 22 21  5
  4  6  3  8 13 12 16 19 19 30 13 25 41 33 31 45 34 18 31 34 20  4
         11 19 21 16 15 28 15 10 31 29 43 46 35 34 30 18 19  3
          3 18 24 20 23 31 17 13 29 11 31 38 35 16 12  7 11  5
          3 24 34 21 23 25 31 32 25 18 31 27 17  4
          3 25 37 28 22 24 32 32 22 22 27 17
          3 24 30 20 16 17 22 36 21 15 17  3
          3 12 17 15 18 18 13 26 20 17  7
             5  7 13 21 15 19 25 15  5
                15 24  9 14 18 22  5
                13 27 11  5 12 25  8
                         (a)
```

Fig. 7. Computer printout; (a) cell data on HeLa cell after 4 hours infection, scanned by the UMSP-I.

HP series are 1.5 to 2.0 times greater than in the PM series. Furthermore, except in four cases DSELECT was able to properly classify at least 85% of the cells in the HP series.

COMPARISON OF BIOCHEMICAL EVENTS AND COMPUTER ANALYSIS

Scanning microscopic measurements at 260 nm reflect the presence and distribution of nucleic acids and to a certain extent, in the objects studied, proteins. Application of nucleic acid stoichiometric staining procedures enhances the measurement of nucleic acid in cell preparations. The amount of dye bound to the nucleic acid is directly proportional to the amount of nucleic acid present. Therefore, variations in the measured extinction values reflect changes of the nucleic acid occurring in the cell. Infection of a cell with polio virus leads to a sequence of profound changes in the protein and

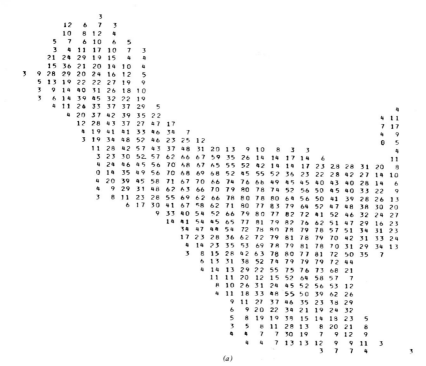

(a)

Fig. 8. Computer printout; (a) cell data on HeLa cell after 4 hours infection, scanned by the UMSP-I.

nucleic acid metabolism of the host, and these modifications should appear as a rearrangement of the structural and/or textural cell images. Such rearrangement has indeed occurred, as evidenced by the changes observed when applying the described filtering techniques to the recorded extinction values.

Visual changes viewed by light microscopy are not evident in a polio-infected HeLa cell culture until a cytopathic effect (CPE) occurs, which greatly alters the morphological structure of the cell. In our experimental system, this effect is extremely pronounced approximately 6 hours after infection. However, intracellular variations in the infected cell can be observed in the ultraviolet light as early as 1 to 2 hours after infection using computer scanning and analysis techniques. These intracellular activities are analyzed by the new algorithms through the sequence of infection until the morphological structural changes begin to dominate. The structural changes are readily distinguished because of the complete disintegration of the infected cells.

The observed intracellular variations of the digitized cell images correlate to the well-known biochemical activities associated with the polio virus multiplications in HeLa cells. This suggests that the computer-aided approach can be used to derive statistical information about biochemical events in the sequence of virus infection. If a specific biochemical process is identified by the appropriate staining methods, it should be possible to study its kinetics and its significance in the virus–host relationship.

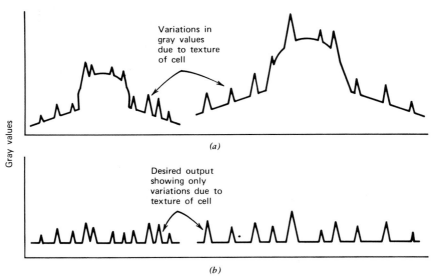

Fig. 9. Conceptual diagram demonstrating the use of digital filters to remove cell structures not relevant to virus infection. (*a*) Before, (*b*) after filter (ideal case).

The computer analysis applied here permits separation of noninfected and infected cells at each hour of infection. However, it is impossible with the existing programs to classify a given infected cell as belonging to a specific hour of infection. Computer algorithms that dynamically adjust to the kinetics of virus infection might be one solution to this problem. Comparison of the results of these algorithms would then lead to the correct recognition and classification of the infected cell.

```
                      2 11   6   5
                  6 17 13  17  18 12
              8  8  2   5   5   5 10
          1 11 13  1       4  10  1
         14 20            7  11
         16  9  7  5          8 16
          1 17  1 17 11      2  9 13  7
          5     2  7      3 11  9  4
         10     6  7  1   8      18 10  6
          1     3  8          2    13  3
    18    6        9       7 15 15 13  8 18  7  1
  7  6  1  4  1  1      8 10 10  8 10  8  5    14  6
  8     5  6             6  0  1  7  4  5  2        7        2
          6             4  9  1  1  4  4  7  3  5      16  7
  1  9  1  3            5  4  4     4  6  4  6  6          12  4
3 3  9  3  0            8  2  5  1        2  5 11          10  2
5 14  8  2            6  3  5        1  2  7  9           9  4
2 16 18 10            8  6  6     2     4 12  2      3  5 13
   8 18 10      7  8  7  6  2     1  3 11      3  7  4  5
   5  6  6      4  5  9  7  4  3  4  9        4  3  6  6
   4 10 18       11  7  3  2  3  3          11      10
   3 10 17   18     9  7  3  1 12  6     3  1  7 10 13
   2  8  7 3 18        7     5 17 21  3  2     6  3  8
   3 10   9  7  1     5 12  2      13     7      13
          7 13   5 11  3              4  1  9 17 15
          3  5  11 15 17  2        5     6  4  7
          8 18  2  0 15 12        2     7  5  1 13
          4 10  0 17 13 13  0  9  7  7  2  6  2 17  5
          7 14  5 15  0  4  7     2  3  3  13
          4  7 12 13 10  8  7        2 16  8
          1     3 10  4  2  4 10  3  7  3  2 13  4
          1 16     8  7 10 17 18  4  3  5  9  6  1
                2     2  6 10  8  9  1  9  5  6
                      3  4  3  2  5  4  4  2  1
```

(b)

Fig. 6. Computer printout; (b) positive difference filter output of the uninfected HeLa cell in (a).

The results obtained here justify a cautious optimism that image scanning techniques may play a useful role in the analysis of trendal cell changes after virus infection. They also very clearly bring out points of concern.

First, the results make it evident that for studies of this type the number of cells sampled at each time interval should be in the order of magnitude of 100 to 150 cells.* The estimates of the standard deviations for some of the image features clearly are affected by the small sample size used in this preliminary study, and the high coefficients of variation indicate that the sample size may have been too small to approximate normal distributions. The obtained average likelihoods for correct classification of individual cells are small, and do not exceed the amounts which could arise spuriously from random fluctuations in the multivariate data sets to an extent which would

*Faster, fully automated, interactive scanning equipment now available make such sample size feasible.

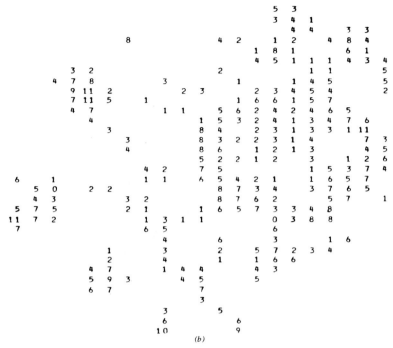

Fig. 7. Computer printout; (*b*) positive difference filter output of cell in (*a*).

make the significance of the results very obvious. With the expected differences between cell feature distributions in normal, and infected cells being that small, a substantially increased sample size would permit a clearer separation of the experimentally attained discrimination from spurious information.

On the other hand, the question at issue in most cases here would not be the identification of individual cells as infected, or not infected, but rather an objective characterization of a trend in cell parameters, as a function of time after infection. The use of the programs to find, and evaluate suitable parameters is an altogether different goal, and presents good prospects. For such a trend parameter the dimensionality problem is far less serious, even though sample size requirements remain high.

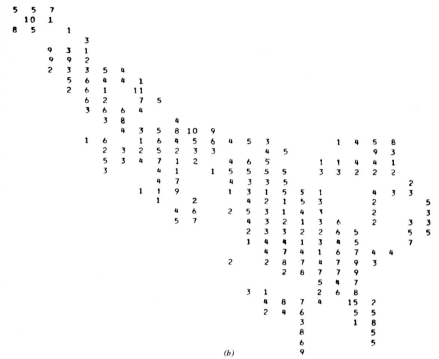

Fig. 8. Computer printout; (b) positive difference filter output of cell in (a).

```
0  0  0        0  1  0        1  1  0        1  1  1        1  1  1        1  1  1
0  X  0        0  X  0        0  X  0        0  X  0        0  X  1        1  X  1
0  0  0        0  0  0        0  0  0        0  0  0        0  0  0        0  0  0
   1              2              3              4              5              6


1  1  1        1  1  1        1  1  1        1  0  1        1  0  1        1  0  1
1  X  1        1  X  1        1  X  1        0  X  0        0  X  1        0  X  1
1  0  0        1  0  1        1  1  1        0  0  0        0  0  0        0  0  1
   7              8              9              10             11             12


1  0  1        1  0  1        1  0  1        1  0  1        1  0  1        1  0  1
0  X  1        0  X  1        0  X  0        0  X  0        0  X  0        0  X  0
0  1  1        1  1  1        0  0  1        0  1  1        1  1  1        1  0  1
   13             14             15             16             17             18
```

Fig. 10. The 18 modified golay square patterns.

243

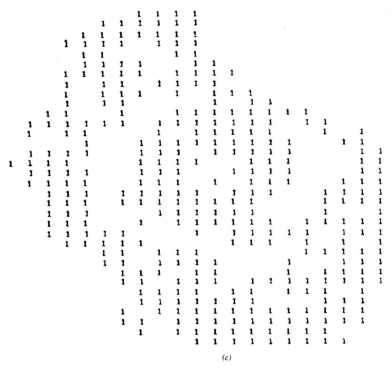

Fig. 6. Computer printout ; (c) binary image of printout in (b).

Table 1 PDF Histogram, Feature 7, PM Series

	Control	5 Hours for Infected
Mean	.0569	.0725
Standard deviation	.0241	.0247
Coefficient of variance	41.7%	33.6%
98% Confidence limits	$.0461 < .0569 < .0877$	$.0614 < .0725 < .0836$

Table 2 PDF Binary Image, Feature 13, PM Series

	Control	Infected for 2 hours
Mean	.0755	.0492
Standard deviation	.0245	.0342
Coefficient of variance	31.9%	68.3%
98% Confidence limits	$.0645 < .0755 < .865$	$.0339 < .0492 < .0646$

Table 3 Linear Combinations of the Selected Features: Control Versus 2, 3, and 4 hours of Infection, PM Series

Hours of Infection							Relative Weights											
	1	2	3	4	5	6	7	8	9	10	11	12	13	14	15	16	17	18
LINCOMB 1: Histograms																		
2		.338	.471			.043			.491		.454	.464		.004	0.10			
3						.405	.376	.167			.698							
4		.227		.356														
LINCOMB 2: Binary images																		
2				.212	.353			.242		.318			.629		.524			
3		.251			.499	.319		.353	.618							.281		
4			.278								.261	.388			.319			.776

246

Table 4 Classification of Cells, PM Series: Control Versus 2, 3, and 4 Hours of Infection

Hours of Infection	% Unclassified Cells		Average Likelihood for Correct Recognition	
	Control	Infected	Control	Infected
LINCOMB 1: PDF Histograms				
2	0.0	0.0	1.8	10.0
3	6.7	13.3	3.2	42.7
4	26.7	33.3	1.5	2.3
LINCOMB 2: PDF Binary images				
2	26.7	36.7	3.1	1.8
3	33.3	20.0	1.4	1.6
4	0.0	0.0	1.9	1.7

Table 5 Final Linear Combination, PM Series: Control Versus 2, 3, and 4 Hours of Infection

Hours of Infection	Relative Weights		Average Likelihood for Correct Recognition	
	LINCOMB 1	LINCOMB 2	Control	Infected
2	0.255	0.697	3.9	1.8
3	0.991	0.132	3.4	63.0
4	0.632	0.775	4.2	3.2

Table 6 Final Linear Combination, Lincomb 3, PM Series

	Control	Infected for 4 hours
Mean	−1.141	.8778
Standard deviation	1.349	1.217
Coefficient of variance	116.2%	136.4%
98% Confidence limits	−1.748 < −1.141 < −.5349	.3306 < .8778 < 1.425

Table 7 Linear Combination of the Selected Features: Control Versus 2, 3, and 4 Hours of Infection, HP Series

Hours of Infection	Relative Weights																	
	1	2	3	4	5	6	7	8	9	10	11	12	13	14	15	16	17	18
LINCOMB 1: Histograms																		
2	.820				.355	.04			.093									
3			.438			.424				.906								
4		.126		.732							.900		.417					
LINCOMB 2: Binary images																		
2		.187		.489	.368	.133			.519			.079						
3	.0031		.009			.814				.313			.003					
4	.443			.285	.339		.467			.157		.603						

Table 8 Classification of Cells, HP Series: Control Versus 2, 3, and 4 Hours of Infection

Hours of Infection	% Unclassified Cells		Average Likelihood for Correct Recognition	
	Control	Infected	Control	Infected
LINCOMB 1: PDF Histograms				
2	0	0	1.4	2.1
3	0	0	10^4	5.2
4	0	0	4.8	3.3
LINCOMB 2: PDF Binary images				
2	17.9	7.1	3.3	9.0
3	10.7	12.9	2.1	31.5
4	3.6	9.7	3.6	6.8

Table 9 Final Linear Combination, HP Series: Control Versus 2, 3, and 4 Hours of Infection

Hours of Infection	Relative Weights		Average Likelihood for Correct Recognition	
	LINCOMB 1	LINCOMB 2	Control	Infected
2	.329	.994	4.4	8.4
3	1.0	.0016	10^4	5.2
4	.44	.898	6.3	47.2

APPENDIX

The 18 binary image features used to analyze the cells are obtained by applying FORTRAN statements at each point in the cell data where the binary image of the PDF output, IEV (IJ) is 1.

```
      I1 = IEV(I − 1, J)
      I2 = IEV(I − 1, J + 1)
      I3 = IEV(I, J + 1)
      I4 = IEV(I + 1, J + 1)
      I5 = IEV(I + 1, J)
      I6 = IEV(I + 1, J − 1)
      I7 = IEV(I, J − 1)
      I8 = IEV(I − 1, J − 1)
      K = I1 + I2 + I3 + I4 + I5 + I6 + I7 + I8 + 1
      IF (K.LE.2. OR. K.GE.8) GO TO 20
      KXOR = XOR(I1, I2) + XOR(I2, I3) + XOR(I3, I4) + XOR(I4, I5)
    1 + XOR(I5, I6) + XOR(I6, I7) + XOR(I7, I8) + XOR(I8, I1)
      IF (KXOR.LE.2) GO TO 20
      K = K + 7
      IF (KXOR.LE.4) GO TO 20
      K = K + 4
      IF (KXOR.LE.6) GO TO 20
      K = 18
      IF (KXOR.LE.8) GO TO 20
    C ERROR STOP
   20 FV(K) = FV(K) + 1
```

Variables I1–I8 form a register that when left shifted end-around and XORed with the unshifted register, gives information about how many $0 \rightarrow 1$ transitions occur in the adjacent 8 points. The array FV(K) is the resulting set of features vectors.

One criterion for selecting these 18 features was the computational simplicity, therefore the computational speed, of extracting rough information about the adjacent points. As a comparison, similar computations on a hexagonal grid would extract only 13 of 16 features described in References 6 and 7, which required a special computer.

Acknowledgments

This study was supported in part by the Deutsche Forschungsgemeinschaft, Sonderforschungsbereich 33, Göttingen, and by Stiftung Volkswagenwerk.

The cheerful and patient cooperation and the excellent technical assistance of Miss Elske Kopp is gratefully acknowledged.

REFERENCES

1. P. H. Bartels, G. F. Bahr, J. C. Bellamy, M. Bibbo, D. L. Richards, and G. L. Wied, A Self-Learning Computer Program for Cell Recognition, *Acta Cytol.* **14**, 486-494 (1970).

2. P. H. Bartels, G. F. Bahr, M. Bibbo, and G. L. Wied, Objective Cell Image Analysis, *J. Histochem. Cytochem.*, **20**, 239 (1972).

3. P. H. Bartels, G. F. Bahr, J. Bellamy, P. K. Bhattacharya, M. Bibbo, and G. L. Wied, Automated Cytodiagnosis, in *Proceedings, Fourth Tutorial on Clinical Cytology*, Chicago, 1971, pp. 1–73.

4. J. C. Bellamy, Computer Recognition of Digitized Image Patterns, Ph.D. thesis Department of Electrical engineering, University of Arizona, Tucson, 1971.

5. A. D. Deitch, A Method for the Cytophotometric Estimation of Nucleic Acids Using Methylene Blue, *J. Histochem. Cytochem.*, **12**, 451–161 (1964).

6. M. Ingram, and K. Preston, Automatic Analysis of Blood Cells, *Sci. Amer.*, **223**, November 1970.

7. K. Preston, Feature Extraction by Golay Hexagonal Pattern Transforms, *IEEE Trans. Comput.*; **C-20**, No. 9 (1971).

8. A. Rosenfeld, Connectivity in Digital Pictures, *J. Assoc. Comput. Mach.*, **17**, No. 1 (1970).

9. H. Selzer, The Use of Computers to Improve Biomedical Image Quality, *Proceedings: Fall Joint Computer Conference*, 1968.

10. ter V. Meulen, P. H. Bartels, G. F. Bahr, M. Bibbo, N. Cremer, E. H. Lennette, and G. L. Wied, Computer-Assisted Analysis of a Carrier Culture Infected with Moloney Leukemia Virus, *Acta Cytol.*, **16**, No. 5 (1972).

11. G. L. Wied, P. H. Bartels, G. F. Bahr, and D. γ. Oldfield, Taxonomic Intracellular Analytic System (TICAS) for Cell Identification, *Acta Cytol.*, **12**, 180–204 (1968).

ADVANCES IN EPIDEMIOLOGICAL SURVEILLANCE

Computer-Aided Monitoring, Surveillance, and Pattern Recognition

RICHARD MOORE

INTRODUCTION

Let us begin by considering briefly the meaning of the terms "monitoring," "surveillance," and "pattern." *Chamber's Twentieth Century Dictionary* defines them as follows:

Monitor	a reminder or admonition; a warning.
Surveillance	a spy-like watching; vigilant supervision; superintendence.
Pattern	[as applicable in this context] a particular disposition of forms or colours; a design or figure repeated indefinitely.

In studying the possible application of computers to medical microbiology, it became apparent to me that by monitoring, by surveillance, and by recognizing patterns of data, modern, third-generation computers could *assist* microbiologists in many ways. The following applications seem to be particularly relevant to current practices in medical microbiology:

1. Recording basic patient-identifying data.
2. Recording data obtained from various processes performed on specimens.
3. Providing recall facilities for any information held in any single record or group of records.
4. Providing totaling and analyzing facilities on any single part or group of parts of all or specifically selected types of records.
5. Providing reports that are required at predetermined times (e.g., day-books, monthly statements, weekly analyses of cross infection).
6. Interrogating of records to determine any significant change from the anticipated norm for specifically selected tests.
7. Indicating an increase, or decrease, on the anticipated norm for pathogenic organisms.
8. Enabling ad hoc inquiries to be made on any single data field or group of data fields within a record file.

No doubt many other applications will emerge in the course of computing experience in this field.

COMPUTING DISCUSSED

Before going further I should like to dispose of some misconceptions about computers. A computer does not possess intelligence surpassing any yet known; there is no "superbrain"; nor, in my opinion, will computers "control the world." The computer, together with various peripheral devices, is an inanimate collection of electronic components which are totally incapable of performing any function unless instructed to do so. After a power supply is connected, very detailed instructions are required before anything can happen at all. The device has the merit of being able to perform the many functions it is instructed to do at a very high speed, and it can repeat these functions with extreme accuracy if instructed correctly. My aim is to utilize this capability in a manner suited to the needs of microbiologists. The methods I should like to discuss are designed to assist in a microbiological sense; no consideration has been given to the costing, billing, and similar financial aspects already performed by some systems, although these facilities may be included if required.

The present methods used to secure clinically relevant information from the analysis of microbiological specimens generate directly and indirectly large amounts of informative details. These details may be considered under two broad headings: (a) determination, and (b) method of determination.

"Determination" is the stage in the acquisition of information about a microbe that leads the trained observer to reach a specific conclusion about the identity of the organism. If considered in association with other pertinent observations and results, the determination may indicate patterns of behavior, frequency of occurrence, locations of outbreaks of infection, and other characteristics of use in epidemiological research.

"Method of determination" is the recording of all tests performed on a specimen, as well as the results of such tests that were pertinent to the conclusion reached. It provides a comprehensive quality control surveillance that may act as a measure of the efficacy of the various tests applied and of the interpretation of such tests. It is desirable that the most complete detail be recorded. By doing so, a "textual image" can be created of the methods and materials in laboratory procedures on which an assessment was made. The information recorded should relate to all suspect organisms as well as to organisms whose significance can be decided only after identifying tests have been completed, and their results are known.

WHAT IS A COMPUTER?

Let us look in detail at a basic "computer." It is comprised of a central processing unit together with many types of peripheral unit, such as a punched card reader, a punched paper tape reader, a magnetic tape unit, an exchangable disk store, a line printer, and numerous other devices, all working on the periphery of the central processor.

A program, possibly comprising many hundred instructions, is loaded into the central processor, and dependent on the requirements of the system, the appropriate part of the "computer" performs whatever task is required at each predetermined stage of the program. After processing is completed, the resultant detail can be recorded on some form of storage device and retained for further use. Each uniquely identifiable collection of information (a record) is written (magnetically transcribed) to an allocated area of the file (a collection of records).

Various manufacturers have different methods of operation for their processing units. Some machines are known as "byte" machines, others as "word" machines. Each term indicates the structure of the unit within the central processor that is available for manipulation by the program. Each term is said to comprise so many "bits" (*binary digits*). For our purposes, the ICL 1900 series of computers is considered, and argument is based on a "24-bit" word. However, the facilities described are applicable to any other make of machine.

Each record consists of a series of "words." Within certain technical constraints there is no limitation to the size of a record. In practice, however, large records are not to be encouraged; rather, a series of smaller, subject-related record files makes the system more manageable. Figure 1 shows how detail can be stored in a record, either as characters (at 4 characters per word) or numerically as a binary representation of the required number. In one word (24 bits) any number up to 8,388,607 can be stored. Each character written magnetically has an octal (#) number equivalent, stored as 2×3-bit patterns per character. For example, the letter A = octal 41 = bit pattern (octal) 100 001 = binary 100001 = numerically interpreted binary 33. This apparently confusing array will be seen to have great use for our purposes.

The structure of the record file requires very careful consideration. The records comprise both patient-identifying and general information, and clinical detail. Many data areas, or fields, hold information of use for such purposes as administration—statistics on populations and locations, and so on. For a system applied nationally there may be a need to identify records that refer to a specific regional authority, hospital, or laboratory. Within these subdivisions of the main file there may be further subdivisions required for, say, types of specimens dealt with by a particular laboratory department. A possible structure for a typical system appears in Fig. 2.

As with any manual filing system, the computer record file must be held in one main key sequence—for example, alphabetical or numerical order. This order may be called the lowest unique identifier. Sometimes it is necessary to have one record area combining with another to describe a

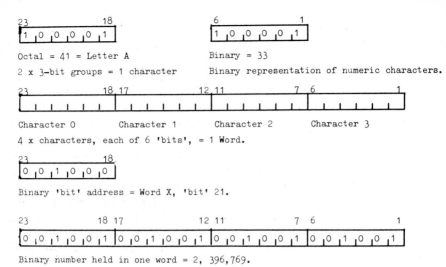

Fig. 1. Forms of storing detail in computer file.

unique identifier. For example, suppose that a laboratory accession number (LAN) allocated to a specimen received by laboratory A from a sequence range 1 to 999,999, might be confused with a number from laboratory B and another from laboratory C, which have similar sequence ranges. In such an event it would be permissible to combine the A of laboratory A with the appropriate LAN to create the unique identifier. Should similar circumstances occur at any other level within the structured system, the level identifier would have to be included in the unique identifier.

After establishing a key sequence for the records on the file, it is usual, but not necessarily vital, to place the identifier fields at the beginning of the record. Groups of fields are then derived from common-data areas (see Fig. 3). Again, it is usual to place the fields in some logical order, but this is not absolutely necessary. By constructing the records with common information in the first part of the record, it is possible to obtain much in the way of nonspecific information for epidemiology and similar purposes. Such grouping is illustrated in Fig. 4. The next segment of the record can be areas having clinical detail pertinent to the species type but also having a common statistical value (e.g., gram stain determination, types of media inoculated, and antimicrobial drug sensitivity information). With a (possibly coded) numeric label for both tests and organism identities (e.g., 076 = *Staph. aureus*; 231 = coagulase test), plus other relevant detail, the last segment of the record would contain details concerning the particular type of specimen—

COUNTRY	STATE	AREA CODE	HOSPITAL IDENT NO	LABORATORY IDENT NO	DEPT	SPECIES TYPE	LABORATORY ACCESS. NO	PATIENT'S NAME		DATE OF BIRTH	SEX
*			**	***			****				
U.K.	MIDD	01	0133	0131	MICR	BACT	143269	SMITH	JS	260727	M
U.K.	MIDD	01	C133	0131	MICR	BACT	143270	BROWN	HS	010150	M
U.K.	MIDD	01	0135	0131	MICR	BACT	143271	WILLIAMS	FG	300671	F
U.K.	SUSS	02	0136	0131	MICR	BACT	143272	JACKSON	HH	281133	M
U.K.	DORS	13	0139	0132	MICR	BACT	001427	ALEXANDER	BP	310355	F
U.K.	DORS	13	0139	0132	MICR	SERO	001428	JOHNSON	JJ	301145	M
U.K.	DORS	13	0139	0132	MICR	VIRO	001429	CLARKE	AC	030946	F
+	++			+++			++++				
ZAMB	LUSA	01	0866	0131	MICR	BACT	000986	UMBWAMA	WC	040845	M

Within entire file: * = 1st Unique Identifier field

      ** = 2nd Unique Identifier field

      *** = 3rd Unique Identifier field

      **** = 4th Unique Identifier field

For localised enquiry: +, ++, +++, ++++.

Fig. 2. Typical computer file structure.

261

Lab. No.	Name		Address	Sex	Hospital	Ward	Specimen	Test No.	Test No.	Test No. Test
178965	Smith	JP	Horsham Sussex	M	0165	A3	Sputa	175	186	187
178966	Brown	AN	Horsham Sussex	F	0165	B2	Faeces	076	092	095
178967	Parkinson	NP	Crawley Sussex	F	0165	B4	Urine	063	071	144
178968	Ponsonby–Smythe	GD	Littlehampton	M	0165	A2	CSF	136	165	033
178969	Jackson	WW	Brighton Sussex	F	0172	A2	HVS	042		
178970	Smith	HN	Hove Sussex	M	0172	B3	Blood	096	097	098
178971	Williamson	WW	Dunsfold Sussex	M	0172	B3	Blood	096	097	098
178972	Churchill	AZ	Mere Sussex	F	0174	A1	HVS	042		
178973	Harvey	SJ	Ruston Sussex	M	0174	B4	Sputa	175	186	187
178974	Singh	RA	Brighton Sussex	M	0177	C2	Blood	096	097	098
178975	Benson	JJ	Capel Sussex	F	0177	D2	W/Swab	143	099	081

Fig. 3. Field grouping for a typical laboratory-based program.

L.A.N.	Name	Address	Hospital	Ward	Specimen	Amp.	Peni.	Clox.	Sulph.	TestA	TestB	TestC
L.A.N.	Name	Address	Hospital	Ward	Specimen	Amp.	Peni.	Clox.	Sulph.	TestA	TestB	TestC
L.A.N.	Name	Address	Hospital	Ward	Specimen	Amp.	Peni.	Clox.	Sulph.	TestD	TestE	TestF
L.A.N.	Name	Address	Hospital	Ward	Specimen	Amp.	Peni.	Clox.	Sulph.	TestD	TestE	TestF
L.A.N.	Name	Address	Hospital	Ward	Specimen	Amp.	Peni.	Clox.	Sulph.	TestG	TestH	TestJ
L.A.N.	Name	Address	Hospital	Ward	Specimen	Amp.	Peni.	Clox.	Sulph.	TestK	TestL	TestM
L.A.N.	Name	Address	Hospital	Ward	Specimen	Amp.	Peni.	Clox.	Sulph.	TestA	TestB	TestC
L.A.N.	Name	Address	Hospital	Ward	Specimen	Amp.	Peni.	Clox.	Sulph.	TestX	TestY	TestZ
L.A.N.	Name	Address	Hospital	Ward	Specimen	Amp.	Peni.	Clox.	Sulph.	TestM	TestN	TestO
L.A.N.	Name	Address	Hospital	Ward	Specimen	Amp.	Peni.	Clox.	Sulph.	TestA	TestB	TestC
L.A.N.	Name	Address	Hospital	Ward	Specimen	Amp.	Peni.	Clox.	Sulph.	TestM	TestN	TestO

Fig. 4. Sample presentation of nonspecific information on computer records.

for example, a quantified (+, + +, + + +, + + + +) microscopic assessment of yeast, mucus, epithelial cells, and others seen in a urine specimen.

Having constructed the record format or formats suited to the required data-retrieval and reporting needs of the system, there remain two major requirements. First, getting the information into the system, and, second, getting it out. Many factors combine to determine the answers to these problems, and each factor must be carefully considered. In determining what detail is to be input and how this is to be achieved, it is first necessary to determine what output is required and in what format. In general, output can be classified as follows:

1. *Mandatory*—that which the system requires to be output at regular intervals and containing predetermined groups of details in a predetermined format.

2. *Optional*—that output deriving from any ad hoc interrogation of the file, or predetermined, occasional interrogation of the file.

For mandatory reports the very formats required determine some of the details that must be recorded in the files. For optional reports, especially those of an ad hoc nature, the difficulty lies in deciding what, if anything, to omit. It is easy enough to say that we do not require this or that detail at present, but can we be sure that such data will not be important at some future time? Ideally, therefore, all known information should be recorded. To achieve this requires the development of a microrecording technique; otherwise the record would be too expensive in terms of recording media, processing time, and computer core-storage utilization. A method of overcoming this is discussed later.

WHAT CONSTITUTES "DETAIL"?

Although it may be accepted that in general, all details should be recorded, it is worthwhile to consider in greater depth what constitutes "detail." If a series of diagnostic tests is performed on a particular organism, and, a specific conclusion is reached from the results obtained (e.g., when a hemolytic streptococcus is suspected but hydrolysis of esculin indicates a fecal streptococcus), it may be argued that to call the organism a fecal streptococcus implies that an esculin test was performed and that there was hydrolysis. If such an assumption is valid, one must consider whether it is necessary to record that an esculin test was performed or that the result was positive. If there is a requirement to know how many esculin tests were performed and whether the results were positive or negative, that fact must be recorded. If only the number of positive tests performed is required, it may suffice to total the number of fecal streptococci recorded. Similar situations may exist in many other instances, and each case must be decided on its merits. When these and other considerations have been determined, and the record format finalized, it will be seen that the information falls into two distinct types:

1. Information, supplied with the specimens, that contains patient-identifying data (name, address, age, sex) plus the type of specimen (urine, feces, sputa).

2. Information obtained as a result of laboratory functions.

Usually there is a time gap between the receipt of type-1 data and type-2 data, and this gap is useful in that it allows us to allocate an area of the record file to the specific LAN, to record the initial data in the record, and, when these functions have been completed, to produce a "day-book" listing. When the laboratory findings are complete, the data can be posted to the appropriate record in the file and the production of any required reports can begin.

Sometimes there is also a requirement to supply an interim report, pending the completion of a series of tests. This need can be met by outputting a print line stating "Further report pending." The method of achieving this is discussed later.

Having determined the required output(s), the method of input must be decided. As stated earlier, the information falls into two groups, called here "prime" and "secondary" data. If it is accepted that on completing the laboratory processes, the required output should be available at a time suited to the recipient's need, (e.g., in time for hospital ward rounds) and that timeliness is preferred to speed, the method of inputting "secondary" data will probably differ from that for "prime" data. Whereas "prime" data can be input at a (comparatively) leisurely rate, allowing normal methods of data preparation, such as punched cards or paper tape, the time factors involved usually require a faster method of input of "secondary" data. This can create many problems and can evoke argument, partly philosophical, about which method is more suited to the circumstances.

One problem associated with the inputting of "secondary" data highlights some of the difficulties involved in considering the type of input hardware—namely, having completed the laboratory processes, the output so determined stands as assessed by a professionally responsible person. Yet any such person may make an error of judgment in interpreting the detail. Regardless of whether the person errs, it is totally unacceptable that an error by a member of the data preparation staff be permitted to enter the record files.

The need therefore, is, for a method that will ensure that the professionally responsible person's findings are correctly transcribed within a predetermined period of time, reducing to a minimum or completely eliminating the intervention of a third party. Many methods were considered and rejected on the grounds of their lacking speed or accuracy, or being incompatible with laboratory functions, space requirements, and similar conditions. Our final choice of input hardware for "secondary" data, considered to satisfy most of the requirements and to overcome most of the objections, is optical mark reading (OMR). In this method we feed into the mark reader a preprinted form (similar to Fig. 5) on which a series of short, black marks has been made in the appropriate positions; the subsequent output from the marked form is transferred to the main-frame computer. The "bit" pattern from the 20 positions on each line are posted, by a file-amending program segment, to a reciprocal area within the appropriate record. In this way the mark made by the technically competent worker

SPUTUM					ZN				LJ	URINE	PREG. TEST		URINE PROTEIN				
M	MP	P	BL	SO	O	+	++	+,+,+	LJ	BL	PT	+	–	O	+	++	+,+,+	7

C.S.F.			SPERMATOZOA						STOOLS/RECTAL SWAB							
									Colour			Composition			Containing		
BL	TU	CL	XN	+	++	+,+,+	NFS	B	Y	G	L	F	BL	MUS	PUS	MF	FG	8

RBC, WBC, CAST AND LYMPHOCYTE COUNTS

< 1 PER CMM			1-10 PER CMM			11-50 PER CMM			> 50 PER CMM				JENNER-GIEMSA			
RBC	WBC	LYM	RBC	WBC	LYM	RBC	WBC	LYM	RBC	WBC	LYM		80	20	8	2	9
H	C	G	H	C	G	H	C	G	H	C	G		40	10	4	1	10

POLYMORPHS %

ACTUAL PARASITES OBSERVED OF:								CYSTS OR OVA OBSERVED OF:								
Ad	Al	Ec	Eh	Gl	Ov	Tv	NTV	Ad	Al	Ec	Eh	Gl	Ov	80	20	8	2	11
Sc	Ss	Ta	To	Th	Tt	No	Sc	Ss	Ta	To	Th	Tt	No	40	10	4	1	12

EOSINOPHILS %

GRAM

Few/Scanty			Moderate (Numbers)				Many/Numerous							
+B	–B	+C	–C	+B	–B	+C	–C	+B	–B	+C	–C	NO	NV	80	20	8	2	13
D	EC	FH	MC	D	EC	FH	MC	D	EC	FH	MC			40	10	4	1	14
PC	SP	VO	Y	PC	SP	VO	Y	PC	SP	VO	Y					15
1	2	4	5	6	7	8	9	10	12					80	20	8	2	16
3	+	++	+,+,+		11	+	++	+,+,+						40	10	4	1	17

MONONUCLEAR %

LYMPHOCYTES %

ORGANISM 1	GENUS								SPECIES/TYPE							QUANTITY			
8	4	2	1	8	4	2	1	8	4	2	1	8	4	2	1	+	++	+,+,+	18
ORGANISM 2																			
8	4	2	1	8	4	2	1	8	4	2	1	8	4	2	1	+	++	+,+,+	19
ORGANISM 3																			
8	4	2	1	8	4	2	1	8	4	2	1	8	4	2	1	+	++	+,+,+	20
ORGANISM 4																			
8	4	2	1	8	4	2	1	8	4	2	1	8	4	2	1	+	++	+,+,+	21

AMPICILLIN   NALIDIXIC ACID

Fig. 5. Typical OMR form.

initiates the transfer of data to the main file without the intervention of a third party, other than the person who actually loads the document into the mark reader.

Mention was made earlier of the way in which a computer can hold data, either as characters (alpha- or numeric) or binary. When the input method is being considered, we must bear in mind certain facts directly related to the hardware capability. For instance, an OMR cannot economically "read" and convert a mark into an alphabetic character, although numeric can be input in a binary mode. Any alpha characters must be input with some other form of hardware. Other data statements are input by answering questions in a "yes/no" situation (e.g., Was the organism sensitive to ampicillin?). A mark in a box containing an "S" indicates sensitivity, and a mark in a box containing an "R" indicates resistance. Absence of either mark in either box would indicate that the test was not performed. Logical grouping of associated data (e.g., antimicrobial drug sensitivities) would permit the transfer of large amounts of data in very few operational stages within the computer program. One line of 20 "yes/no" situations can be transferred as a binary pattern held in one word-area of program core store to one word area of the file record.

The method selected for input of "prime" data is a mixture of punched card and OMR. OMR is chosen whenever an option (yes/no, +ve/−ve, etc.) can be stated. One can now see how the "bit-setting" function to indicate a statement of a condition can substantially reduce the size of the records.

FILE INTERROGATION

Having satisfied all the basic considerations for the inputting and creation of data records, the microbiologist must be able to interrogate the file and to retrieve data. The usual method is to define a specific need, write a program to satisfy that need, and run the program as often as required. Any variation to the normal output, to meet any changed circumstances, requires that a new program be written or that the present program be altered and reconstructed. It is possible, however, to make a program perform more than one specific function, thus satisfying more than one need, if originally specified.

The program suggested here is one of a type known as data management. Whereas the usual program satisfies one or more needs according to the original specification, data management software is concerned with manipulating a series of word areas in a manner determined by the user at run time. This requires a series of separate program segments to be written and held on magnetic backing store (magnetic tape, EDS, MCF). When a run begins, a parameter-card pack is read into the program, and the instructions contained in the cards direct the main segment of program (held in the computer's central processor) to call in, or overlay, the particular routine that performs the function required. On completion of that function, control reverts to the main program segment until another overlay segment is required. This is schematically illustrated in Fig. 6. In this way the main program segment can perform its common function for all needs, and the more specific functions can be performed by the separate overlay segments. This method gives the program great flexibility. In addition, only a small amount of core storage is required, which allows much greater utilization of the core storage available to the machine for running other programs concurrently with the interrogation program.

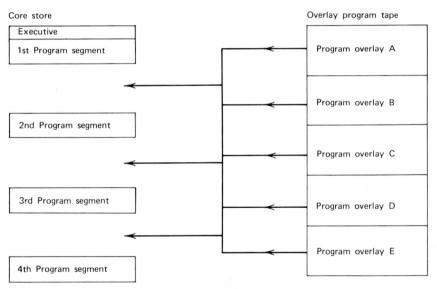

Fig. 6. Schematization of overlay principle of data management program.

The program is endowed with many control functions by means of which the selection of data output can be refined. Certain requirements within any file interrogation can be considered to be key requirements that must be satisfied as a condition for the inclusion of the record in the final output. If any parameter card designated as a key card is not conditionally matched by the related field in the record, the record is rejected. Conditions can be "equal to, greater than, less than, or not." Other non-key requirements can be included in the selection procedures, with an option to reject any record that contains less than a stated number of satisfied parameters (hits), always providing that no key requirements are among the permitted "misses."

The inquiry parameter-card packs, when read into the computer, collectively form a series of binary digits that form a pattern. This pattern is unique. It can extend to one or any combination of fields in the record. Its composition may be determined by the inclusion of, or exclusion of, any chosen fields. Perhaps a digital readout from, say, an image-analyzing computer such as the Quantimet range of products, has been entered into part of the record. This can be reproduced at will, on demand, and it constitutes what is normally understood to be pattern recognition. When such a readout is combined with, say, details of antimicrobial drug sensitivities (possibly with three strengths of the drug), from specific age groupings, by sex, from a particular town or district, occurring between certain dates, and so on, a much more accurate pattern can of course be determined.

Where there is need for surveillance of specimens to detect a change from the anticiapted norm or to observe a certain state or condition, a previously prepared card pack, defining the parameters to be satisfied, can be run against the file at regular intervals. A failure to satisfy the conditional parameters would result in no printout. Detection of change from the specified parameters would yield a printout giving the data to identify the specimen containing the change. This would indicate that a situation has developed that requires investigation—for example, if a change has been detected, it may be due to a change in the organism characteristics, in the methods used, in the media standards, and so on.

When known conditions are to be monitored, similarly, a prepared parameter card pack is run against the file and the existing condition is reported. For example, suppose that an outbreak of infection has occurred in a particular area and regular reporting of the situation is called for. With a surveillance-type program, as many or as few conditions as required can be checked for a general or a highly detailed query, as the circumstances dictate. All monitor and surveillance functions would be performed at predetermined times and frequencies.

For the production of most routine reports on specimens, for any analyses, and for ad hoc inquiries, the output is in tabular form. This is because the interrogation program is highly flexible, and its format must be determined at run time. For the production of reports on specimens, whether for hospital ward reports designed to fit into a Patient's Notes folder or for reporting findings to general practitioners, a separate program, producing an output in a specifically designed format, is required. One function of this program is to indicate that further testing is pending and that the report is only interim when the "final" "bit" is not set. In this case a suitable text is entered to the print line.

When a laboratory has the good fortune to possess its own computer and is working as a single unit, the problem of access to a computer does not arise. When the laboratory is remote from the main-frame computer, or when it is part of a group of laboratories regionally, nationally, or internationally subscribing to a system, the transmittal facilities must meet all requirements of input and outout.

By using modem (*mo*dulating and *dem*odulating) equipment coupled to the hardware in both the laboratory and the computer room, normal telephone systems can be employed to convey data to and from the computer. One simply dials the number of the laboratory or computer room, operates the necessary switches to establish the connection, and transmitting of data can begin. Figure 7 schematically illustrates this method.

To reduce the delay in the event of breakdown in the transmission system, an option to output data from the OMR direct to a paper tape punch can be included in the system. The documents are read, and data are punched into paper tape, which can be transmitted when the line is restored (or taken to the computer room and input direct). The only delay is that during the repair (if the breakdown is slight) or while transporting the paper tape between the two locations. Considerable savings on line charges can be made by using this option in normal practice. Gathering data over time and transmitting in one long run, rather than a series of short runs, avoids holding the line open for long periods when it may be that little or no traffic is passing.

Similarly, a breakdown of the communication link for output from the file can be overcome by outputting a paper tape at the computer, transmitting it when the line is restored, or transporting it to the remote site, and, using the local Modem control unit in "local" mode, feeding it into the printer by way of the paper-tape reader.

Security of the physical components of the file, as well as the information contained therein, must be considered. By taking a copy of the file after each updating, or at specified times each day, and retaining the amendment-data tape for all subsequent amendments, it is possible to ensure speedy restoration of the file in the event of mishap. Additional precautions can include the retention of the last three amendment tapes and file copies on a "grandfather, father, and son" basis, with the latest copy of the file (the new "son") being written over the previous "grandfather" tape. Each tape then moves up one place in the hierarchy.

Normal practices for the physical safety of the files are reasonably easily

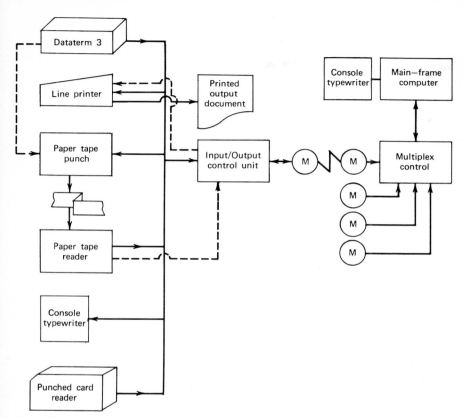

Fig. 7. "Modem" schematic; dashed lines indicate alternative input and/or output.

implemented, but those for the security of the data are more complex. Complaints are increasing with regard to the intrusion of the computer into one's personal life, and it is therefore necessary to provide as much security as possible for medical detail, which is always confidential. There are many ways of achieving security up to the stage of producing the printout, after which problems of security remain as they are with present methods of data handling. The right of access to particular areas of data in the computer can be controlled to a degree by having an "authorized users' file" listing the identity of the user, his current "password" (changed after each separate access), and his access level. An "access level" is applied to each item of the file, and a user can obtain data only from an area that has a level of access equal to or lower than his own (see Fig. 8).

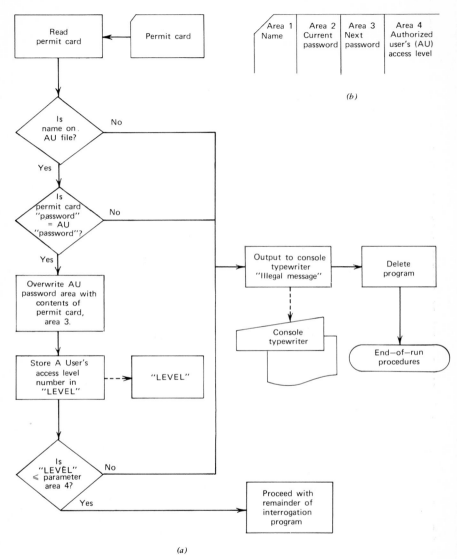

Fig. 8. Schematized access control system: (*a*) flow chart of interrogator validation function; (*b*) permit card layout.

Data can be made even more .secure by using a coding segment of program that will render the data unintelligible on the file medium. An unauthorized user, having a program that will interrogate the file for file-manipulation purposes (generally provided by the manufacturers for legitimate reasons) would only be shown rubbish. The interrogation program, if satisfied with the identity and integrity of the user, would use a reciprocal segment to decode the data and present normally intelligible printout.

When used skillfully, the computer is a powerful ally. This Chapter has indicated ways in which such power can be harnessed for improved data handling in microbiological laboratories.

GLOSSARY

First Mentioned

PAGE 257. **Third-generation:**   describing computers now in use that have computing facilities within the central processor's core store (a collection of ferrite "cores" that can be magnetized or demagnetized to order).

PAGE 257 . **Record:**   an area of recording medium on which information concerning one uniquely identified item is stored.

PAGE 257 . **Record File:**   a collection of subject-related records held on one separately addressable area of recording medium.

PAGE 258. **System:**   the method(s) devised to enable the computer to perform the functions required of it.

PAGE 259. **Program:**   a descriptive noun applied to: (*a*) the act of writing the instructions that control the functions of the computer; and (*b*) the completed list of instructions that form the software for the computer.

PAGE 259. **Bits:**   the shortened form of "*b*inary digi*ts*," which is the term for an electric pulse recorded on a magnetic recording medium or one ferrite core.

PAGE 259. **Octal:**   in composition, eight. The method used in computing to identify to the "executive" (the control program) the character or symbol required. Usually comprises two groups of three binary digits, giving a numerical range from 0 to 7 per group.

PAGE 257. **Data Area or Field:**   that uniquely addressable portion of magnetic recording medium or section of allocated core storage to which magnetic representations of data are transcribed.

PAGE 260. **Laboratory Accession Number (LAN):**   number allocated to a specimen received by the laboratory which is next in sequence within the range of numbers allocated for identifying purposes.

PAGE 264. **Ad Hoc File Interrogation:**   the facility of interrogating the file in any way that is required, which can only be achieved with a program having great flexibility.

PAGE 264.   **Microrecording:** the method of using the presence or absence of an anticipated magnetic condition of a "bit" to convey specific information.

PAGE 265.   **Allocation:** the procedures by which the parts of a file or record are apportioned by "housekeeping" software.

PAGE 257.   **Data:** a neologism, common in the United States as both plural and singular of datum; the standard term in computing for facts given from which others may be inferred.

PAGE 265.   **Outputting:** a term used to convey that the procedures stated are the result of using specific "output" routines in software.

PAGE 264.   **Input:** the term used to convey that the procedures stated are the result of using specific "input" routines in software.

PAGE 265.   **Data-Preparation Staff:** those persons engaged in the processing of information to enable the data to be input in a manner suited to the computer's input facilities.

PAGE 267.   **Alpha or Numeric Characters:** symbols that fall within a predetermined octal number range within the 64-character code that is commonly used in third-generation computers.

PAGE 270.   **Run-Time Option:** the term used to differentiate between the predetermined option of procedure and that procedure required presently as a result of circumstances pertaining.

PAGE 268.   **Parameter Card:** a punched card containing data by which the program determined the limits within which all functions to be performed will be carried out.

PAGE 268.   **Overlay:** a method of core storage utilization; an allocated segment of core store is used many times for different subroutines. On completion of a subroutine, the main program will reallocate the core-sotre to the next subroutine that is required. The next subroutine is then transcribed from backing store into the program segment.

PAGE 270.   **Main Frame Computer:** term differentiating between many smaller types of computer and the large, third-generation machines.

PAGE 270.   **Modem:** equipment by which the electronic pulses are modulated into a form that is acceptable, say, to the post office or telephone company transmission equipment; has the capability of receiving and demodulating signals from an other site.

PAGE 270.   **Local Mode:** the equipment at the laboratory end of the communication link having limited capabilities for working independently from the main-frame computer, the equipment is switched to "local mode" for these functions.

PAGE 271.   **Password:** a phrase or code devised by the person, and known only to him or her, by which the computer program can determine the integrity of the inquirer.

CHAPTER 16

Experiences Within a Computerized National Microbiological Data Retrieval and Analysis System

THOMAS R. NEBLETT

SUMMARY

BAC–DATA is a computerized antimicrobial susceptibility and nosocomial infection reporting service available by subscription to hospitals throughout the United States, presently at an approximate cost of $0.25 per patient medical bed. Isolate data from participating laboratories are submitted on dual mark-sense forms, daily or as completed; by approximately the tenth day of the following month, the laboratory's data are returned, tabulated 10 ways, including trend reports and quarterly plus annual summaries. Over approximately four years, we have found the system to have high value as a laboratory quality control device; it helps us to remain abreast of local and national drug sensitivity changes, and it serves as an existing archive of antibiotic sensitivity data for reference and consultation with our clinical staff. One of the problems we have encountered, which perhaps is shared by other institutions, is the overwhelming amount of data supplied by the system, making digestion and dissemination to others within the hospital a monumental task for the laboratory director or microbiologist. The summary and trend reports that have been recently initiated have supplied a need for simpler and more comprehensive view of the data. Although certain minor problems do exist with the data submitted from individual institutions, we feel that the system's overall content reflects accurately susceptibilities to microorganisms as they exist in the United States.

INTRODUCTION

Some medical institutions in the United States have the benefit of internal computer laboratory data retrieval, which enables them rapidly to analyze their antibiotic sensitivity data and to monitor infections. Such systems have the disadvantage of representing single hospitals, hence no comparisons with data from other institutions are possible. Many hospitals have no data retrieval system at all, and some have thought that a widespread system having the capacity to permit internal and external data comparisons would be valuable to practitioners in the fields of clinical microbiology and infectious diseases. Thus, when we at the Henry Ford Hospital were approached by Dr. David Holvey, then with the Bristol Laboratories Medical Services Division, to become a pilot hospital to help test such a system they had devised, we accepted with enthusiasm.

BAC–DATA SYSTEM

The system that has come to be called BAC–DATA was conceived as a service to the medical profession in which isolated incidence information and antimicrobial susceptibility data sent from individual hospitals to a single computer center would be suitably organized and returned as meaningful reports to the institution for dissemination to its medical staff, infections surveillance personnel, and interested administrators. It overcame the difficulty of certain other cumulative antibiotic data studies extant at that time; information retrieved would be current, because a report would be generated each month.

Pilot hospitals were asked to ascertain value and effectiveness of the system in actual use situations. The institutions selected were located throughout the United States and represented bed size extremes. The Henry Ford Hospital, with 1100 beds and an average daily outpatient load of 2500, was among them.

Input

To submit information to the system, laboratory technologists make a vertical pencil mark in the appropriate space on a printed input data form containing space for two isolates data pertinent to a particular organism.

Names of organisms are represented by two-digit numbers; 17 common isolates are imprinted on the form, with 79 total choices available, plus space for "miscellaneous."

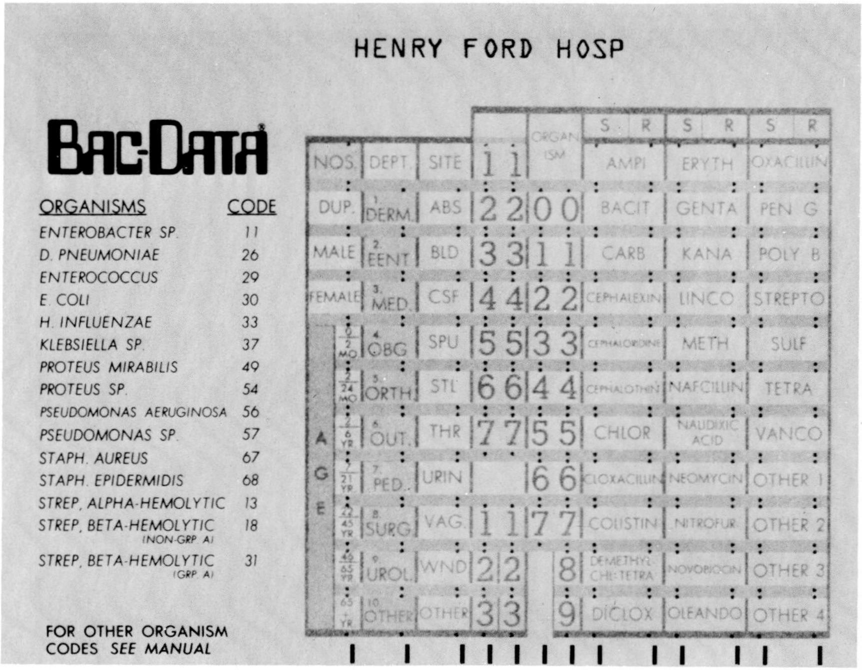

Fig. 1. Mark-sense input form.

Space is also provided for department, body site, sex and age of the patient, and whether the isolate is nosocomial. Antibiotic susceptibilitie. are entered on the card as *S*ensitive or *R*esistant as determined by methodology standards. Under Kirby-Bauer standards, an isolate found to be Intermediate would be entered in the Resistant column. The spectrum of antibiotics contains overlap, but the purpose was to provide information of widest benefit to the greatest number of physicians throughout the country.

Data Return

Data are returned in the following manner. The antibiotic sensitivity report (Fig. 2) includes both local hospital and national data. The national

ANTIBIOTIC SENSITIVITY REPORT

ORGANISMS

	AMPICILLIN	BACITRACIN	CARBENICILLIN	CEPHALEXIN	CEPHALORIDINE	CEPHALOTHIN	CHLORAMPHENICOL	CLOXACILLIN	COLISTIN	DEMETHYLCHLTETRA	DICLOXACILLIN	ERYTHROMYCIN	GENTAMICIN	KANAMYCIN	LINCOMYCIN	METH
E.COLI	9640 78%	85 27%	8287 80%	507 81%	1119 82%	9187 83%	8574 96%	64 48%	7235 95%		75 0%	906 9%	9194 99%	9431 92%	21%	21%
10174	6100 96%	1313 81%	6822 96%	74 10%		223 4%	2648 2%	2732 94%	6177 70%	5737 73%	9539 72%	83 68%				
GROUP A STREP. BETA-HEMOLYTIC	465 96%	39 94%	123 95%	42 97%	66 95%	506 97%	459 99%	22 90%	32 43%		11 90%	515 98%	243 57%	330 15%	429 95%	
1140	16 50%	117 20%	57 94%	39 48%	19 89%	262 91%	516 94%		225 16%	139 13%	505 82%	82 96%				
ENTEROBACTER HAFNIA	66 33%		49 77%			63 41%	63 98%		52 82%		32 93%		65 100%	66 93%		
69	30 93%		31 77%					13 92%	58 77%		65 78%					

2252750040   HENRY FORD HOSP   TOTAL HOSPITAL   DETROIT

ANTIBIOTIC SENSITIVITY REPORT

ORGANISMS

	AMPICILLIN	BACITRACIN	CARBENICILLIN	CEPHALEXIN	CEPHALORIDINE	CEPHALOTHIN	CHLORAMPHENICOL	CLOXACILLIN	COLISTIN	DEMETHYLCHLTETRA	DICLOXACILLIN	TETRACYCLI
ENTEROBACTER SP.	112 11%	19 94%	106 66%	18 55%	61 8%	105 10%	88 94%	1 0%	100 91%	109 85%	4 100%	1 8:
	112			9 11%			1 0%	1 0%	18 83%			
ALCALIGENES SP.	16 93%	2 100%	13 100%	2 0%	5 60%	16 62%	14 85%	3 100%	13 69%	13 92%	1 0%	6
	16			1 100%			3 100%	1 100%				

Fig. 2.

summary has been restricted to include only data being derived from the Kirby-Bauer method, and the format is 8 1/2 × 11 in. (21 × 28 cm), which makes copying for internal distribution quite easy.

Data are expressed as percentage sensitive to a particular drug, with the total numbers of an isolate given along with the numbers tested against the drug. The total hospital portion of the report is a large (28 × 37 cm) format, which is somewhat troublesome for us to handle. On both portions of the sensitivity report new drugs are added periodically as they become available. Number and percentage of organism (Fig. 3) are reported by department and by body site. For each hospital department designated, the uppermost number tabulated opposite an organism is the number found during the previous month. Total isolates for the month is tabulated at the end of each departmental column, and the percentage opposite each name is its fraction of that department's total isolates. This information allows clinical services to be aware of dominant organisms from their patients. The same type of tabulation is supplied for body site origin of isolates. These data enable laboratory and infectious diseases personnel to know dominant isolates from each source and their frequency of occurrence.

Monthly Reports

Monthly trend reports (Fig. 4) contain the 10 most commonly isolated organisms, and these are reported by department, body sites, and drugs against which they were tested. Prior to inception of these reports, data retrieved from the system were so voluminous that we had difficulty handling it, and we had recommended that such a report would be very helpful, to permit quick comparisons with previous months.

Each successive trend report is a cumulative summary of prior reports, and the previous month's document can be discarded. The trend reports' value lies in the user's ability to discern readily any changing values throughout the year.

Bacteriological Trend Report

On the bacteriologic trend report (Fig. 5) one can read numbers of a given organism, isolated by department and throughout the hospital monthly, with quarterly and yearly totals.

Total of all organisms for the month is given, and each identity is expressed as a percentage of the whole.

NUMBER AND PERCENTAGE of ORGANISM

ORGANISM	TOTAL NUMBER	1 DERMATOLOGY	2 E E N T	3 MEDICINE	4 OB/GYN	5 ORTHOPEDICS	6 OUT-PATIENT	7 PEDIATRICS	8 SURGERY
ENTEROBACTER SP.	112 # / %			61 # / 1 %	2 # / 1 %	2 # / 1 %	4 # / 1 %		23 / 2
ALCALIGENES SP.	16 # / %			9 # / 1 %	1 # / 1 %				4 / 1
ALPHA HEMOLYTIC STREPTOCOCCUS	503 # / %	1 # / 4 %	8 # / 7 %	251 # / 3 %	9 # / 2 %	5 # / 4 %	28 # / 4 %	41 # / 4 %	43 / 4
BACILLUS ANTHRACIS	2 # / %				1 # / 1 %	1 # / 1 %			
BETA-HEMOLYTIC STREP NON-GRP A	950 # / %		7 # / 6 %	528 # / 6 %	39 # / 7 %	5 # / 4 %	85 # / 13 %	45 # / 5 %	46 / 4
BORDETELLA PERTUSSIS	1 # / %								1 / 1
BRUCELLA SP.	2 # / %			2 # / 1 %					
CANDIDA ALBICANS	6 # / %			5 # / 1 %				1 # / 1 %	
CITROBACTER GRP E.FREUNDII	233 # / %	1 # / 4 %		121 # / 1 %	2 # / 1 %	4 # / 3 %	9 # / 1 %	45 # / 5 %	11 / 1
CORYNEBACTERIUM DIPHTHERIAE	2 # / %			1 # / 1 %				1 # / 1 %	
D.PNEUMONIAE	368 # / %		3 # / 3 %	165 # / 2 %			23 # / 4 %	126 # / 13 %	8 / 1
EDWARDSIELLA SP	2 # / %			2 # / 1 %					
ENTEROBACTER AEROGENES	154 # / %			95 # / 1 %	3 # / 1 %	2 # / 1 %	2 # / 1 %	3 # / 1 %	11 / 1

Fig. 3.

283

2252750040 HENRY FORD HOSP E.COLI

MONTHLY TREND

PRODUCT		JAN.	FEB.	MAR.	APR.	MAY.
AMPICILLIN		193	299	261	304	343
	%	79	78	86	79	82
BACITRACIN		1			1	1
	%	100			100	0
CARBENICILLIN		189	294	255	298	332
	%	81	79	86	77	82
CEPHALEXIN		1	1	10	62	24
	%	100	100	100	95	100
CEPHALORIDINE		109	208	185	182	275
	%	66	76	89	80	88
CEPHALOTHIN		193	297	258	302	327
	%	74	75	89	82	84
CHLORAMPHENICOL		64	68	97	122	92
	%	98	94	99	96	100
CLOXACILLIN			2		1	
	%		50		100	

Fig. 4.

6–OUT–PATIENT DETROIT MICH

BACTERIOLOGIC-TREND

ORGANISMS		NOV 73	DEC 73	3 mo. Total	12 mo. Total
PSEUDOMONAS SP.	ALL				5
	%				1%
SERRATIA SP.	ALL				1
	%				* %
STAPH. AUREUS	ALL	8	2	18	110
	%	23%	6%	19%	17%

Fig. 5.

The most recent 3-month total plus that for the past 12 months is shown here. This report enables one to determine isolation trends throughout the year and to make comparisons with previous year's isolates. Since these are isolates on which susceptibility tests have been performed, and in our laboratory only dominant or numerous flora receive testing, organisms tabulated in this section very likely represent significant pathogens. Isolates for individual departments are also tabulated. Our practice has been to distribute copies to each department chairman. Thus the microbiologist or pathologist is spared the time and difficulty required to digest and summarize prior to distribution. A very helpful feature of the Bacteriologic Trend report is the numbers of isolates for each month; in addition, under the latest month tabulated on the sheet, we find the grand total of isolates on which susceptibility testing has been performed during the preceding year. The report also contains nosocomial data if reported by the hospital. It

TOTAL HOSPITAL   DETROIT MICH

BACTERIOLOGIC-TREND

ORGANISMS

31/73 PAGE   1

ORGANISMS		3 mo. Total	12 mo. Total
ENTEROBACTER SP	ALL %		
ALCALIGENES SP.	ALL %	16 1%	112 1%
ALPHA HEMOLYTIC STREPTOCOCCUS	ALL %	1 * %	16 * %
BACILLUS ANTHRACIS	ALL %	100 3%	503 3%

Fig. 6.

enables the survey staff to be aware of infections in the same fashion as numbers of organisms.

Antibiotic Susceptibility Summation

An Antibiotic Susceptibility Summation, supplied every 3 months, contains data already supplied to the hospital in other reports but compiled more comprehensively (Fig. 7). It provides on one or two sheets current information on *isolates*, their principal *sources*, what percentage they are of the total, and their susceptibilities expressed to the nearest whole percent to a spectrum of 21 drugs.

We can now ascertain relative site distribution, which is a valuable aid in teaching the frequency and site occurrence of clinical flora; it is especially helpful if one wishes to monitor the frequency of an isolate in the institution. For example, we have been observing an increased frequency of *Serratia* isolates, particularly from respiratory specimens. During the first 3 months of 1973 we found as many as were isolated during all of 1972.

PERIOD JULY 1 TH

ANTIBIOTIC   SUSCEPTIBILITY †

ORGANISM	BODY SITES	NO. OF ISOLATES	% OF TOTAL ISOLATES	AMPI	BACIT	CARB	CEPH	CHLOR	COLISTIN POLY-B	ERYTH	GENTA
CITROBACTER		45	1	29	*	46	31	87	93	*	100
D. PNEUMONIAE	S/T	69	1	98			98	98		98	
	OTH	25	1	100		*	100	100		100	
ENTEROBACTER SP.		31	1	13		55	22	88	77		100
ENTEROCOCCUS	URN	191	4	96		*	29	77	*	74	*
	WND	43	1	95			27	61		57	
	OTH	48	1	96			16	64		66	
E.COLI	URN	588	13	81		78	85	93	93	*	100
	S/T	83	2	85		80	75	99	90	*	99
	OTH	221	5	80		77	79	96	92	*	100
H. INFLUENZAE	S/T	127	3	96			80	100		65	
	OTH	13	0								

Fig. 7.

Copies of this report have been sent to our chairmen of Medicine, Surgery, and Pathology, and to the chairman of the Infections Control Committee. We have also had our print shop duplicate 500 copies for distribution to all who attend medical grand rounds.

Prevalent Pathogens Report

The prevalent pathogens report (Fig. 8) is a quarterly summary of species isolated by body sites in order of their most frequent occurrence. No sensitivity data are included, and the report provides a comprehensive overview of what isolates come from what source. It is duplicated in our institution and will be distributed with the antibiotic susceptibility summary at our grand rounds.

```
REPORT OF PREVALENT PATHOGENS BY BODY SITE AT   HENRY FORD HOSP
                                  2252750040

  1.  THIS REPORT IS BASED ON THE MICROBIOLOGY LABORATORY EXPERIENCE FOR THE
      PERIOD JULY 1 THROUGH SEPT 30 1973

  2.  TO OBTAIN ADDITIONAL INFORMATION CONCERNING THE INCIDENCE OF PATHOGENS
      AND/OR THEIR SUSCEPTIBILITY TO ANTIMICROBIAL AGENTS CONTACT THIS OFFICE.
                                                               876-1040
      ABSCESS         105 ISOLATES           BLOOD          203  ISOLATES
      ---------------------------           ---------------------------
  STAPH.AUREUS                    30 %    E.COLI                     25 %
  PSEUDOMONAS AERUGINOSA          10 %    STAPH.AUREUS               15 %
  STREP.BETA-HEMOLYTIC GRP A       9 %    KLEBSIELLA SP.             10 %
  PROTEUS MIRABILIS                7 %    STAPH.EPIDERMIDIS          10 %
  E.COLI                           7 %    PSEUDOMONAS AERUGINOSA      6 %
  KLEBSIELLA SP.                   7 %    ORGANISM 75                 6 %
  ALL OTHERS                      30 %    ALL OTHERS                 28 %

      C. S. F.        3 ISOLATES           SPUTUM        746  ISOLATES
      ---------------------------           ---------------------------
  NO PREDOMINENT ORGANISM                 PSEUDOMONAS AERUGINOSA     16 %
                                          KLEBSIELLA SP.             12 %
                                          E.COLI                     10 %
                                          H.INFLUENZAE                9 %
                                          STAPH.AUREUS                9 %
                                          H.PARAINFLUENZAE            8 %
                                          ALL OTHERS                 36 %
```

Fig. 8.

SYSTEM IMPROVEMENTS

Recent improvements in the BAC–DATA system include a monthly *Bulletin*, furnishing news and updating information for subscribers to the service. It contains, among other things, items describing how the system is being used in subscriber hospitals.

The BAC–DATA Research Report contains information on the incidence of bacteria and their antimicrobial susceptibilities and investigational information from the field, frequently of a type not usually found in scientific journals.

Within the past year nomenclature used on the reports has been updated to remain abreast of current taxonomic usage. For example, *Aerobacter* was changed to *Enterobacter*. Also, certain organisms that do not fall into the rapidly growing aerobic to facultative category have been eliminated (e.g., *Mycobacterium tuberculosis, Entamoeba histolytica*). Beginning in January 1973, the system initiated nosocomial reporting and recently supplied us with a new card form for reporting that information.

APPLICATIONS OF BAC–DATA SYSTEM

At the Henry Ford Hospital the BAC–DATA reports have been used mainly to aid the microbiology laboratory service. They have been our reference source of local and national sensitivity information and have been helpful many times in telephone discussions of problem situations.

By consulting the reports during the pilot study, our attention was called to a serious error that might have been published. We and our Infectious Diseases staff had assembled data on what seemed to be a most interesting situation—that *Diplococcus* isolates from our laboratory were exhibiting 10 to 12% resistance to tetracycline compared to our awareness of only 1 to 2%. Examination of the sensitivity reports confirmed that the resistance ran about 2% within the other pilot hospitals. We began to suspect that we had erred in identifying *Diplococcus*, rather than assuming that we had a geographically unique resistant strain. Investigation revealed that our technologists had been considering any zone of optochin inhibition around a colony as representing *Pneumococcus*, without further definitive effort. We were reporting a number of *Streptococcus viridans* erroneously, and thorough characterization of the questioned strains failed to confirm most of them as *Diplococcus*.

During the summer of 1970 an increased incidence of an *Enterobacter*-like organism from our blood culture service became evident, with an increase each month. We in the laboratory deal with a high-volume specimen load, and we did not become aware of the trend until we began to isolate a gram-negative specimen that did not clearly fit *Enterobacter*, and with which we were unfamiliar. Several of these were identified by our NCDC as *Herbicola-lathyri* group or as unidentifiable. Late in the year our Infectious Diseases staff conferred with us on the increase, and together we examined the BAC–DATA reports for what was then being called *Aerobacter*; we also checked the "Unknown" organisms column and found a steady increase in isolates since July. The trend prompted an investigation, and we learned that each patient yielding these isolates had received intravenous fluids. We then began examining unopened bottles, and early in 1971 we made the first actual isolation of a small, yellow-pigmented, gram-negative rod from a container that had been presumed to be sterile. Had we been supplied with the trend reports now available, we perhaps would have become aware of the situation sooner.

The BAC–DATA system has aided our laboratory quality control by reflecting changes in internal procedure. We rotate technologists' bench assignments approximately each 4 to 6 weeks; thus any change seen in susceptibility percentages on the next monthly report from a particular specimen category very probably represents a procedural alteration on that bench service. When these changes appear, closer supervisory attention usually brings to light the source of breakdown. On one such occasion our reports began showing susceptibility percentages substantially higher than national figures and we began to check on inoculation techniques. We learned that certain technologists had adopted the unauthorized "time-saving" practice of using the inoculum prepared for our Analytab enteric strips to swab susceptibility test plates, also. The result was too light an inoculum, hence somewhat larger zones and a greater percentage of sensitivity.

We also received comments from members of our clinical staff indicating that we were reporting sensitivities of *Enterobacter* isolates to ampicillin that did not seem to be realistic. On examining our BAC–DATA reports over the previous months, we found that we varied as much as 25% at times from national figures reported for *Enterobacter* ampicillin. This prompted us to reevaluate a modification of the Kirby-Bauer technique we had instituted after what was believed to have been a valid evaluation. We returned to strict Kirby-Bauer practice, and our percentages came back closely in line with national values.

We have recently acquired a Millipore Corporation electronic zone analyzer to read and report our antimicrobial susceptibilities. The installation is not yet completed, but we hope to have the reader generate tape that can be fed to the Fisher-Stevens computer system, eliminating the need for marking manually the data input forms presently in use. This will entail the conversion of Asci code tape onto magnetic tape, because the BAC–DATA system will not read paper tape. It is our intention to reduce labor costs and to enhance reading precision by use of the zone analyzer. Our immediate problem is the additional cost of converting paper tape to magnetic tape.

RESPONSE TO CRITICISMS OF SYSTEM

Certain objections to use of the BAC–DATA system in the United States have been expressed. The dominant one seems to be a suspicion that the data compiled to represent national averages are "tainted" because no assurance exists that the laboratories subscribing to the system are actually performing testing procedures as rigidly as they imply. Moreover, because the subscribers are paying customers, BAC–DATA management may be reluctant to exclude any, even if it becomes known that the quality of the reports of a given institution were not up to standard. A survey of methodology was conducted late in 1972, and the results from 121 hospitals were published in the Second Quarter Research Report, as follows:

1. Mueller-Hinton agar   94%
2. Store agar at 2–8°C   91%
3. Maintain pH at 7.2–7.4   82%
4. Standardize inoculum   82%
5. Inoculate by swab or overlay   93%
6. Use FDA recommended disk potency   90%
7. Use reference organisms: *E. coli*   82%
                            S. aureus   79%
8. Measure zones by caliper, scale, or template   85%
9. Care to discard old plates within 1 to 2 weeks   82%

Serious laxity seemed to exist widely on the following items:

1. Agar depth controlled at 4 mm   28%
2. Daily usage of reference organisms   22%
3. Storage of antibiotic disks in freezer at −14 to −20°C
   until used   11%

Clearly, only slight justification exists for skepticism about the value of the national data, it has been our experience over nearly four years that when we conform as closely as possible to the recommended standard method of testing and reporting using the published Kirby-Bauer zone size criteria, our percentages seldom differ more than 7 to 10% from national averages. Hence, we feel that broadly considered, such criticisms of the system are not justified.

CHAPTER 17

A Data System for Bacteriological Routine and Research

STELLAN BENGTSSON, FRED-OLOF BERGQVIST, WERNER SCHNEIDER

SUMMARY

A data system for bacteriological routine work has been developed which uses punch cards for all patient data contained in the request form and optical mark sheets for registration of the bacteriological diagnosis, antibiotic sensitivity pattern, phage type, and so on.

The system delivers daily lists of all examinations carried out, and twice monthly lists are produced containing all findings of interesting bacteria, such as *S. aureus*, in selected wards in the university hospital during the last month.

The system also forms the basis for a research program for the study of postoperative infections in a new operating ward, where three series of optical mark sheets carry information regarding bacteriological examination of patients, staff, and the air. This information is related to other data on the patient and his operation to produce a condensed version of his treatment period for future statistical analysis.

293

INTRODUCTION

The bacteriological routine laboratory of the Institute of Medical Microbiology at Uppsala University serves as a diagnostic laboratory for a region with approximately 500,000 inhabitants. The region has one central university hospital and seven smaller hospitals. At present about 150,000 analyses are carried out each year; since the laboratory is not connected to the health authorities, all examinations are invoiced.

In 1967–1968 a data system was developed cooperatively by the laboratory and the university data center with the aim of simplifying the processing of answers and invoices to the customers. We also wanted to make possible a statistical and epidemiological analysis of the mass of bacteriological information. This data system, the BACTLAB system, has been in routine use at the laboratory since 1969; it uses ordinary punch cards for the information emanating from the request form and optical mark sense sheets for the results of the bacteriological analysis.

EQUIPMENT

The computer used is an IBM 1130 computer equipped with a 1231 optical mark page reader. The program language used is FORTRAN and 1130 Assembler. A more detailed description of this routine system has been published elsewhere (1). The mark sheet has two sides and contains 1000 possible positions per side (Fig. 1). It contains space for four different bacteria and their respective antibiotic sensitivity patterns, phage types for two aureus strains, as well as less complicated answers such as ordinary throat flora. The form also contains special sections for enteric pathogens and fungi, as well as "free" sections that can be used for special surveys. There is also an open space for annotations—confirmatory tests, tentative diagnoses, and so on. Since no form can have room for the names of all bacteria that may be isolated (e.g., more than 1500 *Salmonellae* species) only the names of the more common pathogens have been included. Should a less common organism be found, a position denoted "kort" (card) is marked, and a punch card with the number of the optical form and the full name of the organism is produced. This results in an answer containing the correct name of the organism.

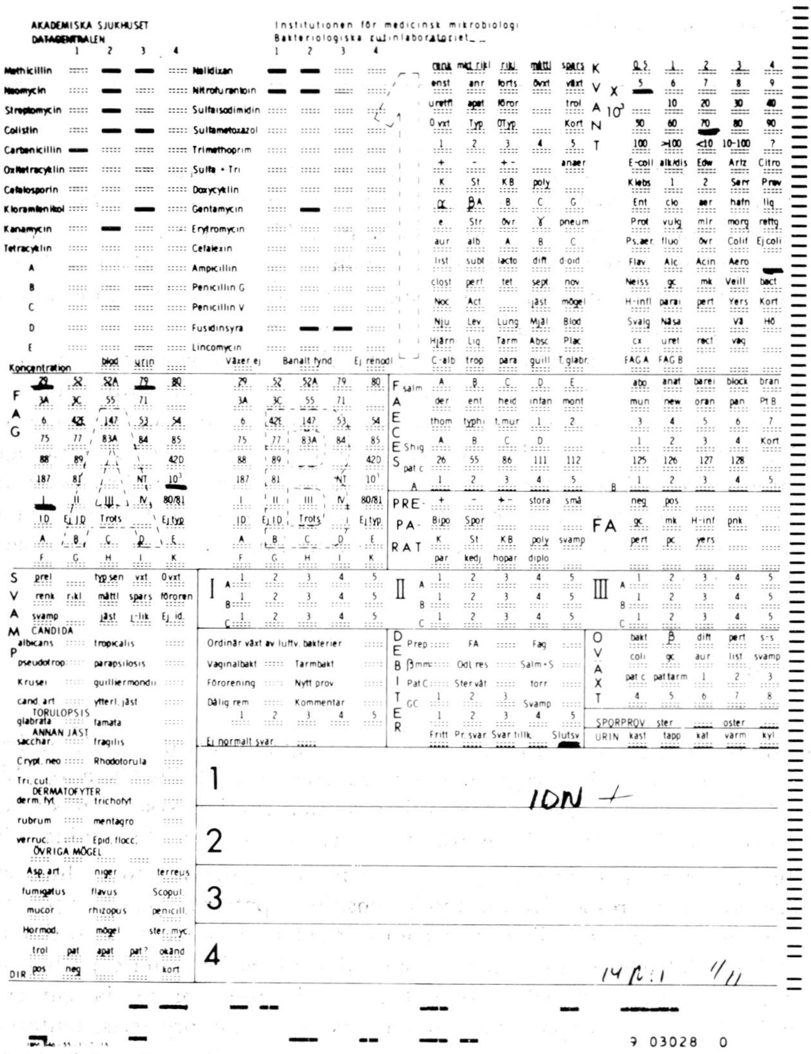

AKADEMISKA SJUKHUSET
DATACENTRALEN
Institutionen för medicinsk mikrobiologi
Bakteriologiska rutinlaboratoriet____

Fig. 1. Optical data sheet of the BACTLAB system. On the reverse side of the form are placed three fields for bacteria similar to field 1.

296

WORKING ORDER

The working order can be described in the following way (Fig. 2). On arrival at the laboratory, the specimen with its request form is unpacked and the form is stamped with a running laboratory number with a paginating stamp. This number is noted on the sample, which is then processed as requested. The data on the request form are transferred to the punch card, which contains all the information regarding the patient, including name, age, hospital, clinic, doctor's name, type of specimen, and desired examination, as well as the laboratory number. The request forms are collected and taken to the laboratory, where they are accessible when the cultures are examined.

The punched cards are sent to the data center, and after control, the information is stored on a disk. Punch cards with errors are returned to the laboratory for correction. The optical mark sheets are delivered prenumbered by the computer (i.e., they are given a number that can be read by the optical reader and they also receive the same number in clear.) This number series is synchronous with the laboratory number series and serves as the connecting link between the patient data on punch card and the bacteriological data on optical sheet.

After one day of incubation, cultures are read, the preliminary diagnosis is noted on the form, and sensitivity tests and complementary confirmatory tests are ordered. After two days the results of the sensitivity tests are marked directly on the optical form. The amount and type of bacteria are also marked by doctor or the assisting laboratory technician according to his instructions. The optical sheets are then transported to the data center for processing. When the optical sheet has been read, the patient information from the punch card with the same number is retrieved from the disk. The information from the stored punch card is now combined with that from the optical sheet to an answer in coded form and stored on disk. These data are printed out with the help of a text dictionary to form the answer from the laboratory.

When negative answers dominate, as in gonorrhoea cultures, the system is simplified: only positive answers use the optical form, and all punch cards lacking an optical sheet counterpart are given negative answers. An optical sheet for serological work has also been developed. This sheet is used in a parallel system with the same principles.

OUTPUT

A list of all answers is produced daily (Fig. 3). It contains the results of all examinations, presented in hospital, ward, and patient age order. This list is

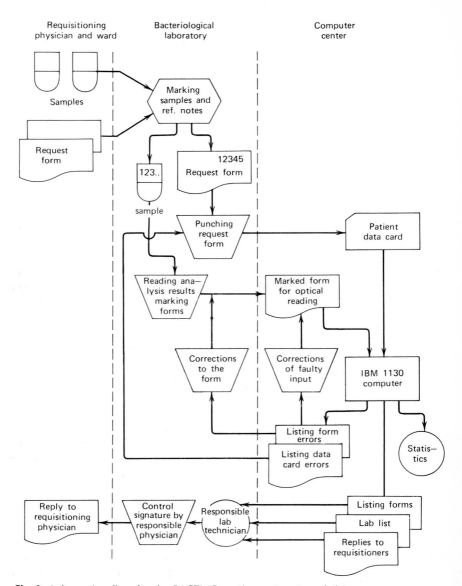

Fig. 2. Information flow for the BACTLAB routine system (run daily).

used to check the work and is also kept as laboratory record (e.g., for answering inquiries). In addition, invoices are produced twice monthly for all customers, hospitals as well as private patients.

Twice monthly lists are produced containing the findings of epidemiological interests from certain selected wards during the last month (Fig. 4). The lists contain patient data, bacteriological diagnosis, antibiotic sensitivity pattern, and phage types for *S. aureus*. The types of bacteria and the wards to be contained in the lists are preselected by means of a special program. These lists are used to control the epidemiological situation in the university hospital: the occurrence of epidemic phage types of *S. aureus* may be followed and changes in the antibiotic sensitivity pattern noted.

A statistics list for monthly reports to the State Bacteriological Laboratory on isolations of gonococci is also produced (Fig. 5). This list contains the number of isolations from male and female patients from the counties served by the laboratory and an antibiotic sensitivity table for the gonococci isolated.

The results of all examinations are saved on magnetic tape for further analysis. This may be done by direct processing of the tape, but it is more convenient to transfer the stored data from the tape to a medical data bank, as is done at the University Hospital in Uppsala. This more general approach allows both direct analysis of the bacteriological data and a study of these in relation to other patient data (2). In this way changes in the sensitivity of bacteria such as *Klebsiella* to various antibiotics can be followed over longer periods. This practice may govern the future use of individual antibiotics in regard to treatment of urinary infections, and so on. The approximate annual cost of the system is at present $25,000 for about 150,000 answers. This includes costs for the forms and the monthly analyses as well as the production of invoices. More than 90% of these costs have been covered by reductions in secretarial staff.

APPLICATIONS

The principles of the BACTLAB system have been used as basis for the collection of data in a large research program concerning the causes of postoperative infections in a recently opened operating ward. In this research program, samples are taken from patients operated on in certain theaters as well as from the staff taking part in the operation. The bacterial contents of the air in the theater is followed using settle plates. Data concerning the patient (e.g., current infections, antibiotich therapy, predisposing factors such as diabetes and malignancy) and his operation (e.g., type

AKADEMISKA SJUKHUSET   750 14 UPPSALA   21.05.73   SIDA   4

```
/POL/        INFEKTIONSKLINIK                                    NNS STS DGE CAP PFL MNS CCO CKK T

421117   ULLSTRÖM ANNSOFID    2 17396   14/5 NÄSS     CARS
•SLUT•
E ANR   AUREUS FAGTYP NT
430822   OLSSON LENA          2 17767   18/5 SVALGS   CARS          2   132 133 3    1
EJ VXT BETAST
550102   ENGLUND E L          2 17748   17/5 SVALGS   RNASJÖ
RIKL    PAPAIN
631214   PETTERSSON S         2 17807   18/5 SVALGS   PAVEK         2   234 222 4
EJ VXT BETAST
640429   ANDERSSON KATRIN     2 17752   17/5 SVALGS   PAVEK
EJ VXT BETAST
640912   ANDERSSON CATRIN     2 17471   14/5 SARS V   CARS
•SLUT•
M RIKL AUREUS FAGTYP SVAG  R 52    53    54    88
CR I   GR III                                                      2   232 122 2    1
M RIKL  BETA A                                                     4   131 111 1
650325   AHLGREN MARIKA       2 17751   17/5 SVALGS PAVEK
EJ VXT BETAST

/82A/        INFEKTIONSKLINIK                                    NNS STS DGE CAP PFL MNS CCO CKK T

450601   HJÄRNER GERT         2 17532   15/5 SARS V   CARS
•SLUT•
RIKL    AUREUS FAGTYP 52      80        GR I  80/81   ID            4   231 133 3    1

/82B/        INFEKTIONSKLINIK                                    NNS STS DGE CAP PFL MNS CCO CKK T

091124   JOHANSSON LENA       2 17777   17/5 SARS V PAVEK
EJ VXT BAKT
190406   GYBO NILS            2 17808   18/5 SVALGS PAVEK
EJ VXT BETAST
ORD VX LUFTVS
```

Fig. 3. List of sorted laboratory answers produced daily and kept as laboratory record.

```
                                           NNS STS DGE CAP PFL MNS CCO CKK T:NX 25 78 3357 4455778888 488  0 IET E   P
                                           AIU MRT OER EMT VUI EET DAX ELA E.TS 92A 90 AC51 62734573458 271 1234 1 DIR TABCDEFGHI

000905  WÅHREN E          04/2
214000  SÅRS V/NÅTTL  AUREUS     2    131 122 2    1                    1                    1       1 1
000905  WÅHREN E          04/9
214567  SÅRS V/RIKL   AUREUS     4    131 132 3    1                    1                    1 1     1 1
000905  WÅHREN E          04/17
215236  SÅRS V/RIKL   AUREUS     2    131 122 2    1                    1                    1 1     1 1
000905  WÅHREN ELISA  04/25
215762  SÅRS V/RIKL   AUREUS     4    131 144 4    1                    1                    1 1     1 1
011204  ANDERSSON AN  04/11
214759  SÅRS V/RIKL   AUREUS     2    131 132 3    1           1      111                   1 1     1
070210  SÖDERBERG ES  04/13
1 6622  URIN /MÅ 100 PYO      442  4 434  44   3 124 43 4
070219  SÖDERBERG ES  04/27
1 8203  URIN / 10  PYO        442  4 434  44   3 434
071204  ANDERSSON AN  04/11
214757  NÄSS /SPARS   AUREUS     2    131 122 2    1               1 1                       1 1 1
111117  OHLSEN HELGE  04/14
214445  SÅRS V/FORTS  AUREUS     2    131 111 1           1 1 1 1           1 1 1
150923  BJÖRKMAN J O  04/2
1 5355  URIN / 50  KL ENT    333  1 434  23   4 434 11 3
151213  BROMAN EVERT  04/2
214001  SÅRS V/RIKL   AUREUS     2    111 111 1    1                    1                    1 1     11
240516  ANDERSSON BR  04/17
416023  FAECES/RIKL   AUREUS     1    111 111 1   14                    1                    1 1     1 1
401217  SPERBER GUNN  04/16
214953  NÄSS /ENST   AUREUS      2    131 111 1    1                    1 1                  1 1     1 1
490527  NYSTRÖM SOLV  04/19
215437  NÄS SV/MÅTTL  AUREUS     2    111 111 1    1                                        1
880921  CARLSSON KAR  04/18
215518  NÄSS /MÅTTL   AUREUS     2    111 111 1    1
880921  CARLSSON KAR  04/24
1 7931  URIN /MÅ 100 KL ENT   331  1 334  23                                                 1
960818  ELWIN BORIN   04/25
215620  SÅRS V/E ANR  AUREUS     2    131 122 2    1                                        1
```

Fig. 4. List of epidemiologically interesting findings produced twice monthly and covering certain bacteria (e.g., *S. aureus*) found in certain wards (e.g., orthopedics) during the preceding month. The list contains the antibiotic sensitivity pattern and *S. aureus* phage types.

301

```
        PATIENT HISTORY FOR      ANDERSSON, ANDERS
        340804-           P103   Prosp Man Ward 70B

Preop.treatment period 1 day
Date of op 720717 Theatre 11  Type No 4200  Wound class 1
Infected case No  Op No    3  Anaesth.type No 400 Drain 0

Date of op Patient      N-T/S  AUR     111 RTD/77
Date of op Patient      Skin/M AUR     111 3A
Date of op Patient      Perin/0 BET AUR


A040  Norlen            N-T/ord
E029  Risberg G-B       N-T/ M AUR     111 RTD/52/79/80
D022  Björkman M        N-T/ M AUR     211 RTD/29/42E/81
      Theatre   1       SED/TOT   75
                        AUR    1       211 47/77
                        AUR    1       411 NT
      Theatre   2       SED/TOT   95
                        AUR    1       211 RTD/77
      Theatre   3       SED/TOT  100
      Theatre   4       SED/TOT   85


Infection after 3 D  Type Woundinfection Slight and Other
Postop. Abi AMPI     Fever

720225  Wound/ S  AUR 1111111111 3A/71

Postop treatment period   11 days
                          Left hospital 720228
                          Increase in hospital stay 5 D
```

Fig. 7. Example of a condensed patient history containing information regarding such factors as predisposition to infection, type of operation and anesthesia, as well as the results of bacteriological examination of the patient, the staff, and the air in the surgical theater at the time of operation. The resulting type of infection and the result of a wound culture also appear.

operative infection rate. This analysis is performed using an IBM 370/155 computer. It is hoped that this research will eventually produce information that will make it possible to reduce the number of postoperative infections now occurring at a rate of about 5% for all operations.

REFERENCES

1. S. Bengtsson and T. Höglund, *Zenbralbl. Bakteriol Hyg. I Abt. Orig. A*, **220**, 146–155 (1972).
2. W. Schneider, *Proc. Mediś 72, Kansai Inst. Inf. Syst. Osaka*, Osaka 1972.

CHAPTER 18

Mobile Laboratory for Yellow Fever Studies

AKINYELE FABIYI

INTRODUCTION

In 1959 a yellow fever epidemic occurred in Umuahia and Uzuakoli, in the former Eastern Region of Nigeria. Vaccinations of persons in the Uzuakoli Leper Colony, the Uzuakoli Secondary School, and the Queen Elizabeth Hospital, Umuahia, were carried out using the mouse-brain-adapted 17D yellow fever virus. In addition, there was begun an extension of studies (1) partly carried out in Lagos to determine the hypothesis (2,3), that the presence of antibodies to group B arboviruses, which are serologically related to yellow fever, interferes with antibody response to successful yellow fever vaccination with the 17D strain. This study required a second vaccination (as challenge) with the French neurotropic mouse-brain adapted vaccine from the Pasteur Institute, Dakar, and the determination of the absence or presence of viremia in those volunteers so revaccinated.

PROCEDURE

Arbovirus group B antibody reaction was evaluated by complement-fixation test; the viruses used included Zika, Dengue, and West Nile. Table 1 shows the results of a neutralization test on post-17D vaccination sera of subjects whose sera before vaccination showed positive or negative complement-fixing (CF) evidence of prior infection with a group B arbovirus (excluding yellow fever). Note that the CF test revealed recent infection, probably within the last six months. The difference on the rate of conversion after yellow fever vaccination between the groups is very significant: $p < .001$. Thus 15% of those with group B antibody before vaccination failed to respond to 17D yellow fever vaccination.

Table 2 represents the control data. It can be seen that when an arbovirus not belonging to group B was used, the differences observed between the groups are not significant. The 15% of subjects who fall within the negative postvaccination reaction to 17D vaccine category, and who before vaccination showed evidence of group B antibody, formed the basis of the category of those revaccinated with the Dakar neutotropic mouse brain vaccine.

ORIGINAL MOBILE LABORATORY

In this phase of this study the mobile caravan laboratory played a very significant part. The distance between Lagos and Umuahia is about 834 km.

309

Table 1 Yellow Fever Neutralization Test on Post-17D Vaccination Sera of Subjects Whose Sera Before Vaccination Showed Positive or Negative Complement-Fixing Evidence of Prior Infection with a Group B Arthropod-Borne Virus (Excluding Yellow Fever)[a]

Prevaccination Sera Group B Reaction (Test)	Postvaccination Sera Yellow Fever Reaction (Neutralization Test)[b]		
	Positive	Negative	Total
Class I, group negative	151 (96.2)	6 (3.8)	157
Class II, group positive	109 (84.5)	20 (15.5)	129
Totals	260 (90.9)	26 (9.1)	286

[a]$\chi^2 = 11.66$; $p < 0.001$.
[b]Figures in parentheses equal percentage of total in row.

Table 2 Yellow Fever Neutralization Test on Postvaccination Sera of Subjects Whose Sera Before Vaccination Showed by Complement-Fixation Test Positive or Negative Evidence of Prior Infection with Ilesha Virus[a]

Prevaccination Sera Ilesha Virus Reaction (Test)	Postvaccination Sera Yellow Fever Reaction (Neutralization Test)[b]		
	Positive	Negative	Total
Class I, group negative	231 (90.9) +	23 (9.1)	254
Class II, group positive	29 (90.6)	3 (9.4)	32
Total	260 (90.9)	26 (9.1)	286

[a]$\chi^2 = 0.0352$; $p = 0.9$.
[b]Figures in parentheses equal percentage of total in row.

The caravan was equipped with a hand-operated centrifuge, a kerosene-operated refrigerator, a portable stove, and cabinets for storage of laboratory materials (tubes, syringes, needles, boiling pans, cotton, alcohol, etc.). In addition, it had a water tank of about 100-gallon or 450-liter capacity. There was no source of electricity. The mobile laboratory also had an area that accommodated 600 (2–3 week old) albino mice and storage area for mouse feed. The mobile laboratory caravan, which was mounted on two wheels, was pulled with a Landrover.

The caravan was equipped with port holes and air vents, and often we could travel only in the morning or evening because of the heat; even so, there was about 20% loss among the mice. Contributing to the temperature problem was the heat generated by the kerosene refrigerator housed in the mobile laboratory. Despite these handicaps, however, we were able to accomplish the following objectives:

1. To revaccinate subjects who, because of the presence of serologically related antibodies to yellow fever, had demonstrated nonconversion or no evidence of yellow fever antibodies after 17D yellow fever vaccination.

2. To determine whether any of these subjects, after being revaccinated with the French neurotropic yellow fever vaccine, had viremia.

The potency of the vaccine kept in the kerosene refrigerator remained the same as when it was tested in Lagos under ideal conditions (electrically operated refrigerator and air-conditioned laboratory). The testing or titration was carried out on the spot, in the field, immediately after use. Mice were examined daily during our stay at Umuahia under a shady tree.

Viremia from subjects were determined by IC inoculation of whole blood diluted 1:2 in PBS with 0.75% bovine albumin fraction V; mice so inoculated were examined on the spot daily for 10 days.

Results

Table 3 summarizes the results of neutralization tests and gives the average survival time on sera of subjects at Umuahia who did not respond to 17D yellow fever vaccination and who were later revaccinated with the Dakar vaccine. The number of subjects is small: 50% failed to convert after revaccination; one subject showed a good response and others had antibody to yellow fever before revaccination with Dakar vaccine. None of these subjects had any evidence of viremia.

NEW MOBILE LABORATORY

Some of the difficulties encountered with the mobile laboratory have been mentioned. Obviously studies performed under such conditions of heat and

Table 3 Yellow Fever Neutralization Test (YFNT) Serum Titers[a] and AST of Subjects at Umuahia Who Did Not Respond to 17D Yellow Fever Vaccination and Who Were Vaccinated with the Dakar Vaccine

Sera[b]	Result	AST[c]	Sera	Result	AST
531 A	< 1.0	4.5	547 A	< 1.0	4.5
B	< 1.0	7.6	B	< 1.0	6.0
C	2.9	10	C	< 1.0	6.5
D	4.0	10	D	< 1.0	7.8
554 A	< 1.0	4.5	549 A	< 1.0	4.5
B	< 1.0	7.5	B	< 1.0	5.3
C	< 1.0	5.5	C	< 1.0	4.6
D	< 1.0	ND[d]	D	7.9	10
566 A	< 1.0	4.3	655 A	< 1.0	4.6
B	< 1.0	7.7	B	< 1.0	5.8
C	1.0	10	C	< 1.0	6.5
D	6.5	10	D	< 1.0	6.0
585 A	< 1.0	4.3	690 A	< 1.0	5.0
B	< 1.0	6.7	B	< 1.0	6.0
C	1.0	10	C	< 1.0	4.6
D	4.7	10	D	< 1.0	6.0

[a]YFNT represents titers expressed as reciprocals of dilutions.
[b]A is pre-17D yellow fever vaccination; B is post-17D yellow fever vaccination; C is pre-French neutrotropic yellow fever vaccination; D is post-French neutrotropic yellow fever vaccination.
[c]AST is average survival time of mice in test.
[d]Not done.

loss of experimental animals, are difficult. With this in mind, another laboratory was designed—hopefully better, more efficient, more modern, and more versatile. Figures 1 and 2 indicate layout of this modern mobile laboratory, equipped with gas-operated refrigerator, autoclave, electrically operated centrifuge, incubator for tissue culture, hot water heater, three sleeping bunks, and a centrally located work bench. In addition, an air conditioner will be installed. The mouse area has inlet and outlet fans plus storage and mouse disposal bins. The laboratory is designed to sleep three

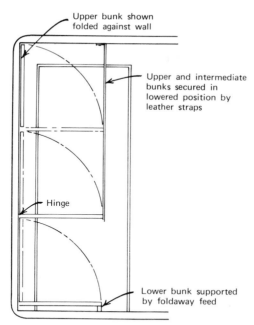

Fig. 1. Cross-sectional view of the mobile laboratory, showing three folding bunks and method of storage.

and possibly four (using the central bench). It can be fully contained in a truck or used as a caravan trailer pulled by a cab; it requires an electric generator, which can be placed behind the caravan and/or pulled on two wheels.

This design is versatile, and such a mobile laboratory can be equipped as a multipurpose clinical and biomedical laboratory. If it is pulled by a cab, the caravan can be made stationary while the cab is being used by other laboratory personnel to carry out surveys elsewhere. Thus for public health purposes collection of specimens can be undertaken by one team while another team performs laboratory investigation on the spot. This laboratory also can be equipped to do serological surveys and other biomedical tests on the spot.

Acknowledgments

The Virus Research Laboratory at the University of Ibadan, Nigeria, is supported in part by the Rockeffeller Foundation, New York. New York.

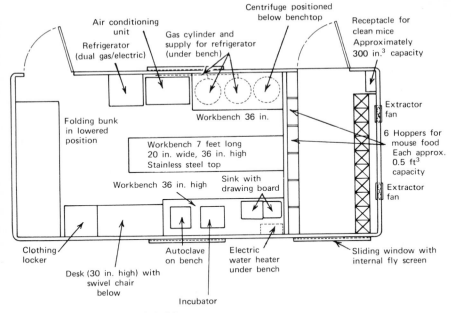

Fig. 2. Plan view of the mobile laboratory.

The drawings of the Mobile Laboratory were made with the collaboration of the Nigeria General Motors Limited.

REFERENCES

1. A. Fabiyi and F. N. MacNamara, The Effect of Heterologous Antibodies on the Serological Conversion Rate after 17D Yellow Fever Vaccination, *Amer. J. Trop. Med. Hyg.*, **11**, 817–821 (1962).

2. P. D. Meers, Yellow Fever Vaccination by Scarification with 17D Strain: An Appreciation of the Present Position, *Trans. Roy. Soc. Trop. Med. Hyg.*, **51**, 338–345 (1957).

3. H. De Roever-Bonnet and J. Hoekstra, Yellow Fever Vaccination by Scarification with 17D Vaccine, *Trop. Geo. Med.*, **10**, 289–291 (1958).

PART D

CURRENT TRENDS IN SIMPLIFIED DIAGNOSTIC TESTS

New Colony Markers Due to Vital Staining During Growth on Dye Containing Agar

V. BONIFAS, G. DEMIERRE, O. RIBEIRO

SUMMARY

Colonies of gram-negative bacilli and gram-positive cocci grown on nutrient agar supplemented with at least one of 31 dyes can be divided arbitrarily into unstained, weakly stained, or strongly stained colonies. Few dyes have specific staining properties—water blue and triphenyltetrazolium chloride stain either one of two separate groups of gram-negative bacilli. Their mixture can be used to isolate *Salmonella*. Congo red produces color patterns in colonies of *Staphylococcus* that permit recognition of serotypes in 80% of seroagglutinable wild strains.

INTRODUCTION

Colonies of bacteria can often be differentiated by their form, structure, color, and degree of opacity. Skill in discriminating among these characteristics is a function of the experience of the observer. Attempts at developing additional colony markers to ease and to speed up reading of isolation plates seemed to be promising. Long-lasting experience has already exemplified the utility of color markers such as those which develop when lactose-fermenting organisms develop on lactose-indicator agar; their colonies are stained red. Colonies of virulent *P. pestis* have been found to accumulate Congo red from the nutrient agar (1) much as they do with hemin (2). More recently, it was observed that *Pseudomonas pseudomallei*, grown on MacConkey agar, grew as red colonies even though it did not ferment lactose (3).

This chapter summarizes experiments involving the growth and staining behavior of 31 bacterial strains on solid media containing at least one of 31 dyes, as well as trials in the routine laboratory. Part of this material has been published in the theses of two of the authors (4, 5).

MATERIALS AND METHODS

Media

Nutrient Base

Bactotryptone 10 g, Bactoagar 16 g, distilled water 1000 ml, pH adjusted to 7.0 before autoclaving.

Supplements

All supplements were added from separately sterilized solutions. D-glucose and other sugars were prepared as a 10% solution in water and added to make a final concentration of 0.2%. NaCl was added to a final 0.05% or 0.2% concentration. The dyes were prepared as 2% solutions in ethanol and were added to the melted agar to a final dye concentration of 0.02%. Dye solutions were freshly prepared before use.

Dyes

Monoazo group:	Chrysoidin Y, methyl red, Sudan II
Diazo group:	Chlorazol black (E), trypan blue, Evans blue, Congo red, benzopurpurin 4B, naphthol Blue-black, Bismarck brown Y
Phenylmethane group:	Victoria blue 4R, Victoria blue B, water blue, night blue
Anthraquinone group:	Alizarin, quinalizarin
Tetrazolium group:	2,3,5 = triphenyltetrazolium chloride (TTC), neotetrazolium
Xanthene group:	Rhodamine B, eosin yellowish, eosin bluish
Quinone imine group:	Gallocyanine, Nile blue sulfate, nigrosin
Miscellaneous:	Hematoxylin, azolitmin, BZL blue, Visba green, Alcian blue, tartrazin, orcein

Some of these compounds were of unknown degree of purity and old; origin is not referred to, since the dyes cannot be purchased from the same producers.

Bacterial Strains

Gram-positive cocci:	*Staphylococcus aureus* (type strains of Pillet), *Streptococcus* type D
Gram-negative bacilli:	Enterobacteriaceae (*Salmonella*, diverse subspecies), *Shigella* (diverse subspecies), *Klebsiella* (diverse species), *Pseudomonas aeruginosa*

All strains were lyophilized. In experiments, they were kept in deep nutrient agar, inoculated with a needle, from which overnight slants were inoculated for each experiment.

EXPERIMENTAL PROCEDURES

Petri dishes containing 15 ml of agar were inoculated with the platinum loop to obtain isolated colonies on a portion of the surface. They were incubated overnight or longer at 37°C. Examination was made through a stereomicroscope with magnifications of 12.5 and 32. Photographs were taken through the stereomicroscope. The following characteristics were recorded: size, outline, surface and color markings of the colonies, and color changes of the surrounding medium.

RESULTS

It was soon apparent that color patterns developed only on media containing D-glucose (sometimes other sugars) but no added salt; all descriptions fit the appearance of colonies on salt-"free," D-glucose-supplemented dye agar.

General Effects of Dyes on Colony Growth and Staining

The many experimental data are summarized in Table 1 under four headings: no 'staining, weak staining, strong staining, and growth inhibition. The inhibitory effect favors gram-positive cocci, particularly *Staphylococcus* (12 out of 31 dyes). The three staining effects are distributed in an apparently haphazard manner among stains and dyes, group coherence being found within gram-negative bacilli in most cases. Within structural groups of dyes, there is evidence that side chains are more important than the basic structure in producing the effect. Within gram-negative bacilli, differentiation in two groups occurs with only two dyes, water blue and TTC (in the staining reaction), and alizarin (in the inhibitory effect). Despite the strong staining effect of a variety of molecules on both gram-negative bacilli and gram-positive cocci, only Congo red produced different staining patterns that were studied further.

Possible Differentiation Between Gram-Negative Bacilli

The differences found between groups of generi of gram-negative bacilli served to divide them in two groups—those which stained blue on water blue agar but did not take up any TTC to reduce it, and the others. Under the circumstances, we used media supplemented with a mixture of water blue and TTC on which confirmation of the preceding observations was obtained. Thus the following generi or species grew on this medium as red-centered colonies with a white margin: *E. coli* indole positive, *Klebsiellae*, *Proteus*, *Pseudomonas aeruginosa*, and *Shigella boydii*, whereas the following grew as a blue colony: *Salmonellae* (including *Arizona* and *Citrobacter*), *Shigella sonnei*, and *E. coli* indole negative. This prompted us to organize a field trial in the routine laboratory.

Table 1 Staining Properties of 31 Dyes on Colonies

Dye	None	Staining Effect[a] Weak	Strong	Growth Inhibition
1. Chrysoidin Y	−	−	G −	G +
2. Methyl red	G + G −	−	−	−
3. Sudan II	G + G −	−	−	−
4. Chlorazol black E	−	G −	G +	−
5. Trypan blue	−	G −	G +	−
6. Evans blue	−	G −	G +	−
7. Benzopurpurin 4B	−	−	G + G −	−
8. Naphthol blue-black	G −	G +	−	−
9. Congo red	−	−	G + G −	−
10. Bismarck brown Y	G +	−	G −	−
11. Victoria blue 4R	G −	−	−	G +
12. Water blue	G −	−	G + G −	−
13. Victoria blue B	G −	−	Stre	G +
14. Night blue	−	G −	Stre	Sta
15. Alizarin	G +	−	G −	G − [b]
16. Quinalizarin	G − , Stre	−	−	Sta
17. TTC	G −	−	G + G −	−
18. Neotetrazolium	−	G −	−	G +
19. Rhodamine B	−	−	G + G −	−
20. Eosin bluish	−	G −	G +	−
21. Eosin yellowish	−	−	G + G −	−
22. Gallocyanin	Stre	G −	−	Sta
23. Nile blue	−	−	G − , Stre	Sta
24. Nigrosin	−	G + G −	−	−
25. Hematoxylin	G − , Stre	−	−	Sta
26. Azolitmin	G −	G +	−	−
27. BZL blue	−	G +	G −	−
28. Visba green	−	−	G −	G +
29. Alcian blue	−	−	G −	G +
30. Tartrazin	G − G +	−	−	−
31. Orcein	−	G −	−	G +

[a]Stre = *Streptococcus* D, Sta = *Staphylococcus*, G + and G − = gram positive and gram negative organisms, respectively.
[b]*Proteus* and *Pseudomonas aeruginosa*.

More than 500 stool specimens were examined by the standard method for *Salmonella* and *Shigella* (Table 2), as well as with an improvised method using as first plating medium, either directly or after enrichment, water blue–TTC agar. We isolated 25 strains of *Salmonella* with both methods, 20 with the standard method, and 18 with the other one. In this trial, 5 times more colonies were tested from water blue–TTC agar than from standard plates because we had to learn that *Citrobacter* and *E. coli* indole negative grew as blue colonies. Nevertheless, 5 cases went undetected with the standard method, whereas 7 escaped us with the improvised method. Table 2 summarizes the results of the field trial, which were reported elsewhere (7).

Table 2 Comparison of Standard *Salmonella–Shigella* **Isolation Method with Water Blue–TTC Agar Isolation Medium: 565 Samples**

	Standard[a]	Water Blue–TTC[b]	Both Methods
Colonies tested	70	341	
Number of *Salmonella* found	20	18	25
Number of *Shigella* found	—	1	
Number of *Salmonella* unseen by either method	5	7	
Petri dishes used	5200	2036	

[a]Samples of stool spread onto MacConkey, brillant green, and DCLS agar either directly or after 24 hours and 96 hours enrichment in selenite broth. Suspected colonies were tested with the universal *Salmonella* phage (6), for urease, fermentations, and antigens.
[b]Samples of stool spread onto water blue–TTC agar, either directly or after 24 hours or 96 hours enrichment in selenite broth. Suspected colonies were spread onto MacConkey agar to eliminate lactose fermenters. Suspected colonies were then tested for fermentations and antigens.

Possible Differentiation Within the Genus Staphylococcus

Staphylococcus colonies are strongly stained when grown on media containing one of 5 of the 6 diazo compounds mentioned in Table 1. Among them, Congo red offered the definite property of producing color markings in colonies of each serotype, determining patterns that made them easily recognizable (Fig. 1). The patterns are due to the distribution of fine or gross precipitates of a dark brown compound unlike the blue insoluble acid form of Congo red. Congo red is known to strongly associate with most carbon compounds; this property was known long ago and was used to diagnose amyloid disease (8, 9). However, stained colonies of *Staphlococcus* can be suspended in water and separated in two components—unstained cells and the precipitate. Thus the compounds associating with Congo red in the colonies are not solid constituents of the cell walls. On the other hand, as in most serotypes, the Congo red binding substance(s) remain within the colonies; in type 9 cultures agar precipitation lines can be observed. Type 9, like types 11, 12, 186, and 218, contains two colony types as revealed on Congo red agar, but not on any other medium. The type 9 reference strain and the one isolate of this type present one weakly and one strongly stained colony type. When colonies of the two types are close together on Congo red agar, as soon as 24 hours of incubation at 37°C has passed, a cooperative effect toward the formation of a precipitation line of a brown precipitate between the two colonies are observed (Fig. 1). Another observation was made on aging cultures grown either on glucose agar or on Congo red agar. Subcolonies which are not stained, appear within stained colonies. The subcolonies can be grown easily, and they were all tested for their antigenic makeup as well as for their capacity to take up Congo red and form the brown compound markings. In all cases, they were as agglutinable as the mother strain; but they never recovered the capacity to form stain patterns. All these observations lead us to believe that the compounds responsible for Congo red combining are entities independent from the antigens. This is exemplified in the results of a field trial (10) summarized below.

The field trial was organized with two aims: to survey the serotypes occurring both in random isolation and in chronically infected patients during a limited period of time (3 months) and to study the behavior of *Staphylococcus* infecting chronic patients submitted to intense antibiotic treatment. Typing with colony markings was effected at the same time. Strains were chosen for their adherence to the laboratory definition of the virulent group (i.e., mannite positive, coagulase positive, hamolytic, and DNAase positive strains). One missing characteristic was allowed. Table 3 presents the results of the trial. Among the agglutinable strains (50%), 10% did exhibit a discordant "morphotype," to use the current term. Also, a number of these strains were seen to possess antigens characteristic of two serotypes—in particular 1 and 2—as well as traits of both morphotypes. In a number of instances the capacity to form the brown compound was more marked than in the type strains. In the nonagglutinable strains, however, 15% corresponded to a known morphotype. Since the cause for nonagglutinability is unknown, interpretation of this occurrence remains obscure.

One important feature of the results was that all chronic patients re-

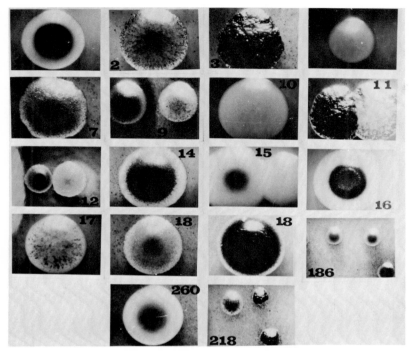

Fig. 1. Compound photograph of the colonies of *Staphylococcus*-serotyping reference strains of Pillet, 24 hours on Congo red. Type 18 is shown after 24 and 48 hours incubation; types 9, 11, 12, 15, 186, and 218 contain two colony types on Congo red agar. From left to right, the types are I, II, III, 6, 7, 9, 10, 11, 12, 14, 15, 16, 17, 18 (twice), 186, 260, and 218.

mained infected with *Staphylococcus* of the same morphotype throughout. Agglutination of the strains isolated each week or at longer intervals was irregular, and some were nonagglutinable despite repeated subcultures in the proper media. This has thrown doubt on the validity of some of the procedures used or at least on the significance of the color markings, whose independence from antigenic makeup may lead to another way of classifying *Staphylococcus* subtypes.

Role of Sugars

Most bacteria used in the experiments eventually ferment glucose, and possibly other sugars. Reduction of tetrazolium to formazan has been viewed as a substitution of tetrazolium as an electron acceptor to natural substrates,

Table 3 Serotyping and Morphotyping of 150 *Staphylococcus* **Isolates**

Serotyping	Agglutinable Strains (%)	Nonagglutinable Strains (%)
Corresponding morphotype	40	None
Discordant morphotype	10	15[a]
New morphotype	None	35
Total	50	50

[a]Known morphotype in nonagglutinable strains.

particularly those derived from sugars, when the latter is lacking (11). This cannot be the case in the present experiments, since *Salmonellae* and *E. coli*, both glucose fermenters, behave in opposite manners with regard to both TTC and water blue. In addition, the change required from *E. coli*—to move from the TTC reducers, water blue unstained colony formers, to the *Salmonellae* type of behavior, TTC nonreducers, and water blue stained colony formers—has to do not with sugar metabolism but with indole production. Hence various attempts were made to substitute glucose with other sugars and to note possible changes of response to dyes. Figure 2 presents type 3 *Staphylococcus* grown on Congo red agar supplemented with glucose, maltose, and fructose 24 hours after incubation at 37°C. Not only

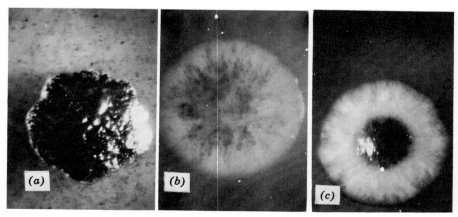

Fig. 2. Type III *Staphylococcus* grown on Congo red agar supplemented with (*a*) glucose, (*b*) maltose, (*c*) fructose.

are the markings different, but the form on the colony is not the same. It is evident here that the roughness of type 3 depends on the combined action of glucose and Congo red. With *Salmonellae* substitution of glucose with other fermentable sugars lead to TTC reduction by many subspecies. These results indicate that metabolic rules in colonies may not be identical to those known from suspension cultures.

Mode of Formation of Patterns in Colonies of Staphylococcus

For patterns to be formed, Congo red must enter the colonies. One can reason either that (*a*) the dye reacts with colony components when they are formed, perhaps at the surface of the solid medium, whereupon the components are lifted upwards by colony growth, or (*b*) the dye invades the growing colony by passive diffusion or by active pumping mechanisms. In deciding between the two major mechanisms (i.e., the reactive mechanism and the invading mechanism), we utilized a transfer of colonies from a simple growth medium to a Congo red medium with help of stainless steel grids, 400 mesh, which were disposed as follows. A large section of grid was sterilized and set at the surface of a thick layer of nutrient agar. A thin layer of nutrient agar was then poured on top of the grid to embed the latter and to form a smooth surface on which *Staphylococcus* could be spread with the loop. After 24 hours of incubation, the grids were seized with forceps and moved to a thick layer of Congo red agar. Further incubation for various lengths of time ensued. During incubation at 37°C, Congo red diffused toward the colonies and invaded them, staining these areas progressively from bottom to top, whereas the dark compound was formed later than penetration. Figure 3, presents two serotypes of *Staphylococcus* after 18 hours of incubation, posttransfer. Incubation at 4°C or on glucose free Congo red media at 37°C served to produce a weak pink staining of the colonies. Hence a diffusion mechanism exists, but another mechanism comes into play when glucose activates it. These results also show that Congo red is not bound to cell surfaces, since it can diffuse through the colonies. This indicates in turn that the cell surface is very peculiar, since Congo red binds cellulose, cellulose acetate, polysaccharides, and even protein. Indeed, these properties constituted a major obstacle in colony transfer experiments and led us to choose stainless steel grids.

DISCUSSION AND CONCLUSIONS

Despite the apparent ease in producing staining patterns of colonies on media containing glucose and dye(s), our understanding of the rules of

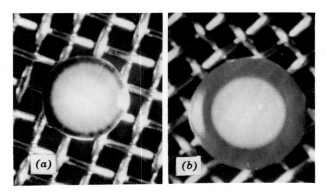

Fig. 3. Two serotypes 24 hours after transfer from growth medium to Congo red medium.

colony life remains scanty. In recent years, new interest in parameters of colony growth has arisen (12, 13). However, the study of the metabolism of bacteria while in colonies might be a difficult task, since there is an apparent heterogeneity of topographical situations for cells in a colony: some are exposed to the source of nutrients at the agar surface, others are exposed to the atmosphere on top, and some are in between, far away from both. It is known that metabolites stream out of colonies and alter the composition of the surrounding medium. One may well assume that circulation of nutrilites also takes place in other directions. The movements of dye molecules may reflect on the functions of the circulatory mechanisms of colonies, and their apparent selective staining activity may be due to physicochemical conditions in each species related to the net charge of the dye molecule in the colonies and the conditions of the subcapillary system within the colonies. Hence one major difference between cell conditions in colonies and in suspension cultures may well be due to the actual concentration of metabolites, which would be much higher in colonies than in suspension cultures in the immediate vicinity of the cells.

The results presented here indicate that color markers may be used in practical work in accordance with practice in the field; however, a better understanding of their physiology is required to make them safe tools for bacteriologists.

REFERENCES

1.   M. J. Surgalla and E. D. Besshey, *Appl. Microbiol.*, **18**, 834–847 (1969).
2.   S. Jackson and T. W. Burrows, *Brit. J. Exp. Pathol.*, **37**, 570–576 (1956).

3. Ch. H. Zierdt, *J. Infect. Dis.*, **125**, 325–326 (1972).

4. O. Ribeiro da Cunha, Développement d'une Nouvelle Technique de Culture Permettant la Différentiation Rapide Manuelle ou Automatique des Sérotypes de *Staphylococcus aureus*, Thesis. University of Lausanne, 1971.

5. G. Demierre, La Coloration Virale des Colonies Bactériennes: Étude de Quelques Colorants, de Quelques Souches Microbiennes et de Quelques Sucres, Thesis University of Lausanne, 1972.

6. H. Fey and A. Margadant, *Zbl. Bakt. Parasitkde. (I. Abt.)*, **218**, 376–389 (1971).

7. P. Mavrothalussitis, G. Demierre, and O. Ribeiro da Cunha, *Pathol. Microbiol.*, **39**, 47–48 (1973).

8. H. Bennhold, *Deutsch. Arch. Klin. Med.*, **32**, 143 (1923).

9. M. G. Stemmermann and D. Auerbach, *Amer. J. Med. Sci.*, **208**, 305 (1944).

10. O. Ribeiro da Cunha, G. Demierre, and V. Bonifas, *Pathol. Microbiol.*, **34**, 8–9 (1973).

11. J. Lederberg, *J. Bacteriol.*, **56**, 695 (1948).

12. S. J. Pirt, *J. Gen. Microbiol.*, **47**, 181 (1967).

13. M. S. Hochberg and J. Folkman, *J. Infect. Dis.*, **126**, 624–635 (1972).

CHAPTER 20

Enzymatic Analysis in Microbiology

H. BRUNNER AND G. HOLZ

INTRODUCTION

The term "enzymatic analysis" in its present sense is generally understood to mean an analytical procedure with the aid of enzymes, including the determination of enzyme activities themselves (1). The advantages and the particular value of enzymatic analysis lie in the properties of the enzymes involved—that is, in their ability to react specifically with the individual components of even complex mixtures, in their effectiveness in small amounts, and in the high sensitivity of the reaction, which takes place at room temperature. Another great advantage of enzymatic analysis is that the procedure can be made very efficient and even automated. These features combine to make enzymatic analysis the optimum method for solving analytical problems in microbiology.

Enzymatic analysis and microbiology have been closely linked ever since the discovery of enzymes. Investigations of the physiology of microorganisms yielded important contributions to the understanding of the course of reactions and reaction chains in microbial metabolism. This knowledge, which is essential for any work on microorganisms, is obtained today mainly by enzymatic analysis and related methods.

ENZYMATIC ANALYSIS AS A TOOL FOR IDENTIFYING MICROORGANISMS

Enzymatic analysis has proved its effectiveness in the identification of microorganisms based on the fact that enzymes and/or metabolites are excreted into the surrounding nutrient medium. The enzyme specificity and activity make it possible to define the microorganisms e.g., Barre (2) and Cato (3) propose enzymatic determination of optically active lactic acids as a means of classifying lactobacilli. However, classic assay procedures are usually qualitative or semiquantitative; they are standard procedures in food technology, especially in the dairy industry, for screening programs for suitable strains and contaminating strains. Tests for proteolytic, rennet, lipolytic, catalase, and special carbohydrate metabolizing activites are conducted (4–7).

In the medical diagnosis of pathogenic bacteria, estimations of fibrinolytic, hemolytic, lecithinase, phosphatase, and amidase activity, are of vital importance (8–10). A combination of serological methods and enzymatic analysis, moreover, represents a great advance in taxonomy. In fermentation, a contamination of pure cultures can be detected by enzymatic analysis

335

much earlier and in an extremely short time compared with any classic microbiological technique. An enzyme activity significant for the unwanted organism can be used as a tracer. E.g. in continuous cultures of Achromobacteriaceae in carbohydrate containing nutrient broths, acetatekinase (AK) activity may not be detectable. The emergence of AK activity may indicate contamination by *E. coli*, which has very high AK activity under these conditions.

ENZYMATIC ANALYSIS IN STUDIES ON METABOLISM OF MICROORGANISMS

An important task in the biochemistry of microorganisms is the elucidation of reaction steps in metabolism. Microorganisms respond to variations in the composition of their nutrient media by changes in their physiological and biochemical properties. The enzyme pattern of the organism can be altered in a specific way and certain metabolic pathways may be influenced so that they proceed predominantly in a predetermined direction. Especially in this field, which may be termed cell engineering, rapid and specific methods of analysis for the determination of metabolites and enzymes activites are essential. Usually enzyme activities lend themselves far more readily to analytical procedures than their substrates, which are often subject to rapid metabolization. For example, analysis of the pathway characteristic enzyme activities can easily determine the distribution of glycolysis or the pentosephosphate cycle or, in the case of *Aspergilli* and *Penicillia*, the participation of the glucose oxidase branch.

ENZYME INDUCTION

Microbial variability can be understood only by taking account of the enzymatic factors. Wortmann (11) found evidence of formation of amylase by certain bacteria growing on starch-containing culture media. The response to alterations in the surrounding milieu—called "adaption" in the early literature (12–16)—was explained by the studies of Monod (17), which led to the development of our present theories on enzyme induction and repression mechanisms. Induction of chains of metabolism by one substrate only was recognized more than a quarter-century ago (18–20).

The investigation of metabolic pathways by simultaneous adaption (18) or by sequential induction of enzymes (21) has been an outstanding aid in the study of metabolism of aromatic substances, such as amino acids and nucleic bases.

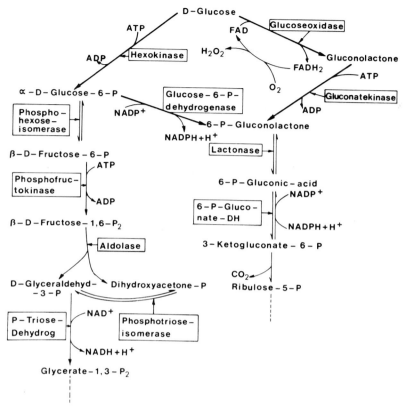

Fig. 1. Branching of pathways in glucose metabolism.

All studies on enzyme induction involve enzymatic analysis. By systematic analysis of ornithine *trans*-carbamylase in *E. coli*, Gorini (22, 23) discovered that cultivation of the organism in the presence of arginine stops the enzyme synthesis, the activity decreasing to 0.01 (see Fig. 2). Arginine does not inhibit the carbamylase itself but acts as a repressor in *de novo* synthesis. The study of microorganism metabolism was stimulated by the development of the continuous cultivation techniques, which provide independence of the changing conditions of batch fermentation (24–26). This method permits the stabilization of a certain state of growth, theoretically for an indefinite time. Thus organisms can be characterized in any growth phase.

The dependence of growth (biomass), carbon source consumption (glucose), and activity of an intracellular proteinase on dilution rate in a study on continuous culture of a strain of *Penicillium chrysogenum* (27) is shown in

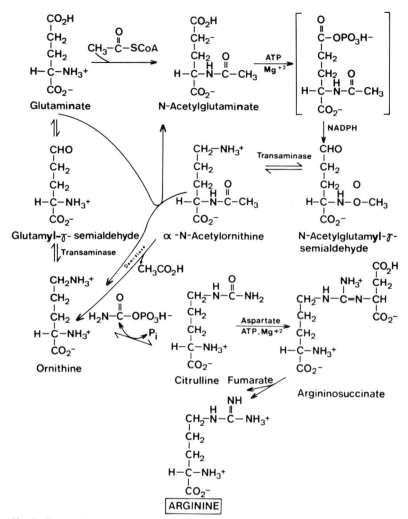

Fig. 2. Biosynthesis of arginine.

Fig. 3. What might be called the superimposed maximum of intracellular proteolytic activity coincides with a change in morphology—increased vacuolization—and, as demonstrated in recent studies with oxygen, becomes the limiting substrate (28). By this means proteinase was defined as an enzyme whose significance resides in the transformations of the intracellular protein pool.

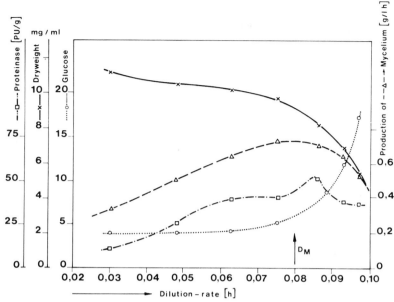

Fig. 3. Continuous culture of *Penicillium chrysogenum*: Dependence of proteinase activity and growth on dilution rate.

FRACTIONATION OF SUBCELLULAR PARTICLES

In the biochemistry of cell organelles the fractionation of subcellular particles such as mitochondria, lysosomes (29), nuclei, and ribosomes, is followed by the analysis of markers (i.e., mainly enzymes specific for these structures) (30). The purity of such preparations is best illustrated by analysis of contaminant enzymes. The following enzymes can be mentioned as markers, cytochrome oxidase and enzymes of CAC as markers for mitochondria; DNA-dependent RNA-polymerase and nicotinate-mononucleotide-adenyl-transferase for nuclei; cathepic proteinases, acid phosphatase, and ribonuclease for lysosomes; and enzymatically active ribosomal proteins, such as protein L11, which shows peptidyl transferase activity, for ribosomes.

METABOLISM INHIBITORS

Metabolism inhibitors act by interfering in enzymatic processes. Inhibition of enzymatic reactions in microorganisms is the basis of the action of preservatives. The inhibition by preservatives of certain steps in glycolysis and in the tricarbonic acid cycle in food-spoiling organisms was demonstrated by Rehm (31–33). Thus enzymes are not themselves inhibited by sulfite, the latter forming adducts with pyridine coenzymes, which are bound (e.g., as NAD-sulfite-compound) to the enzyme protein of a dehydrogenase.

G lyceraldehyde-phosphate-dehydrogenase (GHD) and lactate-dehydrogenase of *E. coli* are strongly inhibited, whereas the GDH of yeast needs high concentrations of sulfite for inhibition. Alcohol-dehydrogenase, an important enzyme for yeasts, is not affected. Sorbic acid inhibits carbohydrate metabolism comparatively strongly. The action of so called antimetabolites can usually be explained by substrate analogy causing competitive inhibition (e.g., malonate for succinate, sulfanilamide vs. *p*-amino benzoic acid, canavanine vs. arginine). As a rule, the action of the numerous antibiotic substances can be traced back to an inhibitory effect on certain enzymes. Resistance of quite a number of microorganisms to pharmaca has been intensely studied by many laboratories. Moyed's review of biochemical resistance to pharmaca (34) is an example of the application of enzymatic analysis in this field.

ENZYMATIC ANALYSIS IN FERMENTATION RESEARCH AND INDUSTRY

The production of substances of microbial origin in the fermentation industry involves extensive control of microbial growth, turnover of nutrient components, and assay of the desired product, including the analysis of unwanted by-products.

To circumvent the risk of leaving the control of a fermentation cycle solely to the personal experience and feeling of the operators, greater efforts were made to explain the factors influencing the process. We now have physical, physicochemical, chemical, biological, and enzymatic methods of analysis for reliable determination, control, and regulation of limiting substrates, toxic substances, and metabolites, their rate of formation and breakdown, and last but not least, the activity of certain enzymes. Enzymatic analysis continues to gain ground in process control in the fermentation industry.

As an example, let us present some aspects of our everyday procedures for the production of microorganisms as a source of enzymes. Among the most important measurements are periodic determinations of the concentration of substrate during fermentation. Determinations of acetate, ethanol, aspartate, citrate, galactose, glucose, glutamate, glycerol, lactate, and other substances, are routinely performed by enzymatic analysis throughout the growth cycle. These data are correlated to certain enzyme activities, the relationship being of great importance in process optimization.

Candida mycoderma grown on glycerol-containing nutrients is used as raw material for the extraction and purification of glycerokinase (GK) (24–35). This enzyme is inductable by its substrate, and if the glycerol concentration decreases below a certain value, the recovered mycelium does not contain sufficient GK. The specific activity (defined as units of activity per milligram of protein) becomes too low to justify purification procedures. The fermentation process is easily controlled by using enzymatic determination of glycerol, which may be performed in a few minutes. Figure 4 illustrates the principle of the enzymatic determination of glycerol.

$$\text{Glycerol} + \text{ATP} \xrightarrow{\text{GK}} \text{3-Glycerol-P} + \text{ADP}$$

$$\text{3-Glycerol-P} + \text{NAD}^+ \xrightarrow{\text{GDH}} \text{Dihydroxyacetone-P} + \text{NADH} + \text{H}^+$$

or

$$\text{Glycerol} + \text{ATP} \xrightarrow{\text{GK}} \text{3-Glycerol-P} + \text{ADP}$$

$$\text{ADP} + \text{PEP} \xrightarrow{\text{PK}} \text{ATP} + \text{Pyruvate}$$

$$\text{Pyruvate} + \text{NADH} + \text{H}^+ \xrightarrow{\text{LDH}} \text{Lactate} + \text{NAD}^+$$

Fig. 4. Principle of the enzymatic analysis of glycerol.

To produce $D(-)$-3-hydroxybutyrate-dehydrogenase (3-HBDH), an enzyme used in clinical biochemistry for the analysis of acetoacetate (1), *Rhodopseudomonas sphaeroides* is fermented on acetate-containing media. The enzyme catalyzes the reaction of 3-hydroxybutyrate to acetoacetate with reduction of NAD^+ (see Fig. 5). In cultivation of the organism for the 3-HBDH, the following reactions must be taken into consideration.

$$\text{acetate} \rightarrow \text{acetyl-CoA}$$

introduction of acetyl-CoA into Krebs cycle, energy gain

$$\text{acetyl-CoA} \rightarrow \text{acetoacetate-CoA} \rightarrow \text{3-hydroxybutyrate-CoA}$$

$$\rightarrow \text{poly-3-hydroxybutyrate}$$

No 3-HBDH is used in these reactions (Fig. 6). As soon as the medium's concentration of the energy-supplying substrate—acetate—is too low to

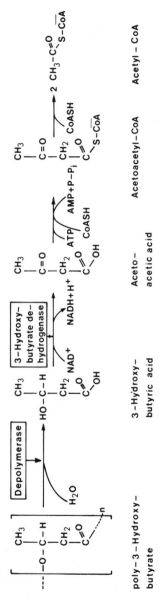

poly-3-Hydroxy-butyrate   3-Hydroxy-butyric acid   Aceto-acetic acid   Acetoacetyl-CoA   Acetyl – CoA

Fig. 5. Turnover of poly-3-hydroxybutyrate in *Bacteria*.

cover the energy requirements of the cell, the accessory polymeric poly-3-hydroxybutyrate is assimilated: 3-hydroxybutyrate→acetoacetate→acetyl-CoA. The reaction of 3-hydroxybutyrate to acetoacetate is catalyzed by 3-HBDH; prior to this 3-HBDH is not formed in adequate amounts (36, 37). The correlation between acetate consumption and enzyme formation is detected by enzymatic analysis, as in Fig. 7. Malate-dehydrogenase (MDH), which may be regarded as an enzyme indicating the activity of the Krebs cycle, thus a parameter of the energy-supplying turnover of acetate, exhibits growth-related activity. MDH is quite similar in its physicochemical properties to 3-HBDH. Therefore, a minimum amount of MDH in relation to HBDH is wanted. By the use of the enzymatic determination of glutamate, which is present in the medium, it was possible to control the activity of the citric acid cycle—that is, to keep down the formation of the unwanted MDH (39).

Summing up, we can conclude that the use of enzymatic analysis in microbiology or any other biological or biochemical field is justified and highly recommended whenever the substances of interest can be assayed only by tedious chemical methods, if at all. Such analysis can also be valuable whenever optical antipodes are involved and, of course, whenever enzyme activities themselves are the subject of interest.

Thus the laborious analytical methods in routine control of the "classic" processes of production of lactic, citric, and gluconic acid could be replaced by the simple and specific methods of enzymatic analysis. As Oppermann pointed out with the expansion of analysis, especially enzymatic analysis, coupled with the increasing application of modern computing methods for simulation and automatic control systems of processes, we are approaching the development of new microbiological feedback loops (i.e., control loops).

REFERENCES

1. H. U. Bergmeyer, *Methods of Enzymatic Analysis*, 2nd enlarged ed., Verlag Chemie, Weinheim, 1973.

2. P. Barre, *Ann. Technol. Agr.*, **15**, 203–209 (1966).

3. E. P. Cato and W. E. C. Moore, *Can. J. Microbiol.*, **11**, 319–324 (1965).

4. S. C. Prescott and C. G. Dunn, *Industrial Microbiology*, McGraw-Hill, New York, 1949.

5. H.-J. Rehm, *Industrielle Mikrobiologie*, Springer-Verlag, Berlin, Heidelberg, New York, 1967.

6. K. J. Demeter, *Bakteriologische Untersuchungsmethoden der Milchwirtschaft*, Ulmer, Stuttgart, 1952.

7. J. Williams, *J. Soc. Dairy Techn.*, **13**, (1), 46 (1960).

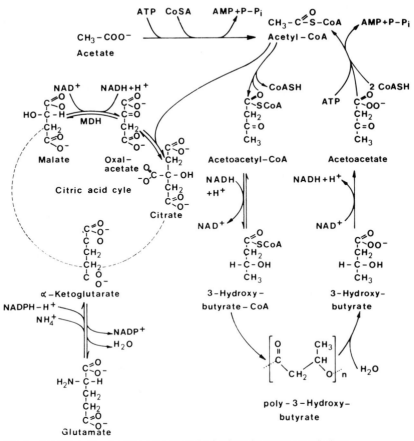

Fig. 6. Reactions involved in acetate and 3-hydroxybutyrate metabolism.

8.   L. Hallmann, *Bakteriologie und Serologie*, Thieme, Stuttgart, 1961.

9.   W. F. Harrigan and M. E. McChance, *Laboratory Methods in Microbiology*, Academic Press, London/New York, 1966.

10. M. Thorney, *J. Appl. Bacteriol.*, **23**, 37 (1960).

11. J. Wortmann, *Z. Physiol. Chem.*, **6**, 287 (1882).

12. H. Karström, *Ergebnisse Enzymforschung*, Vol. 8, Akademia Verlagsgesellschaft, Leipzig, 1938, p. 350.

13. K. Linderström-Lang, *Handbuch der Enzymologie*, Vol. 2, Akademie Verlagsgesellschaft, Leipzig, 1940, p. 1121.

14. J. Monod, *Growth*, S, 223–289 (1947).

15. S. Spiegelmann, in *The Enzymes*, J. B. Sumner and K. Myrbäck, Eds., Vol. I, Part I, Academic Press, New York, 1951, pp. 267–306.

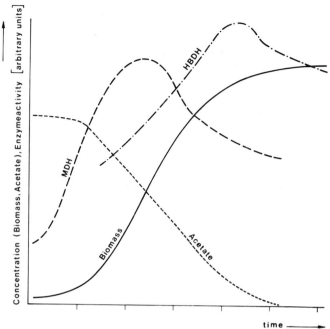

Fig. 7. Distribution of MDH and 3-HBDH during growth cycle of *Rhodopseudomonas sphaeroides*.

16. R. Y. Stanier, *Ann. Rev. Microbiol.*, **5**, 35 (1951).

17. J. Monod, *Recherches sur la Croissance des Cultures Bacteriennes*, Hermann, Paris, 1942.

18. R. Y. Stanier, *J. Bacteriol.*, **54**, 339 (1947).

19. J. L. Karson and H. A. Barker, *J. Biol. Chem.*, **175**, 913 (1948).

20. M. Suda, O. Hayaishi, and Y. Oda, *J. Biochem. (Japan)*, **37**, 355 (1950).

21. M. Cohn, J. Monod, M. R. Pollock, S. Spiegelmann, and R. Y. Stanier, *Nature*, **172**, 1096 (1953).

22. L. Gorini and W. K. Mass. *Biochem. Biophys. Acta*, **25**, 201 (1957).

23. L. Gorini, W. Gunderson, and M. Burger, *Cold Spring Harbor Symp. Quant. Biol.*, **26**, 173 (1961).

24. I. Myers and L. B. Clerk, *J. Gen. Physiol.*, **28**, 103–112 (1944).

25. J. Monod, *Ann. Inst. Pasteur*, **79**, 390–410 (1950).

26. A. Novick and Scilard, *Proc. Nat. Acad. Sci. (U.S.)*, **36**, 708–719, (1950).

27. H. Brunner and M. Röhr, *Appl. Microbiol.*, **24**, 521–523 (1972).

28. H. Brunner and M. Röhr, *J. Gen. Microbiol.*, in press.

29. Ph. Matile and A. Wiemken, *Arch. Mikrobiol.*, **56**, 148–155 (1967).

30. G. D. Birnie, *Subcellular Components, Preparation and Fractionation*, Butterworths, London, 1972.

31. H.-J. Rehm, *Zentralbl. Bakteriol. II*, **121**, 491 (1967).

32. P. Wallnoefer and H.-J. Rehm, *Z. Lebensm. Unters. Forsch.*, **127**, 195 (1965).

33. H.-J. Rehm and J. Baltes, *Z. Lebensm. Unters. Forsch.*, **137**, 295 (1968).

34. H. S. Moyed, *Ann. Rev. Microbiol.*, **18**, 347 (1964).

35. H. U. Bergmeyer, G. Holz, E. M. Kauder, H. Moellering, and O. Wieland, *Biochem. Z.*, **223**, 471 (1961).

36. C. F. Boehringer, German patent 1238422, 1967.

37. D. H. Williamson, J. Mellanby, and H. A. Krebs, *Biochem. J.*, **82**, 90 (1962).

38. G. Holz, *Zentralbl. Bakteriol. I*, Suppl. 2 (1966).

39. G. Holz, in H. U. Bergmeyer, Ed., *Methoden der enzymatischen Analyse*, Verlag Chemie, Weinheim, 1970.

CHAPTER 21

Miniaturized Microbiological Techniques for Rapid Characterization of Bacteria

DANIEL Y. C. FUNG AND PAUL A. HARTMAN

SUMMARY

Many routine microbiological tests involving both liquid and solid media have been miniaturized by utilizing the Microtiter system, which was originally developed for serological studies. In this paper we describe two multipoint inoculation devices, methods for gas detection in mini-fermentation tubes, mass transfer of reagents into growth cultures for mini-IMViC tests, detection of motility, and biochemical tests in mini-semisolid media and mini-agar slants.

Using capillary tubes, semiquantitative detection of catalase activity by isolated colonies was achieved. Also, a capillary-tube procedure was developed for reading various reactions in mini-litmus milk tubes.

These techniques require about 5% of the material and 10% of the time needed for conventional procedures in studies involving many cultures; the techniques can be adopted to other procedures, and they are easy to use.

INTRODUCTION

Rapid and miniaturized microbiological techniques usually increase the efficiency of routine microbiological work by saving space, material, labor, and time. The concepts and applications of a variety of miniaturized microbiological techniques have been described elsewhere (1–4.).

In pursuing miniaturization of routine microbiological techniques, several general aspects must be considered. First, suitable culture vessels must be designed to accommodate small volumes. Plastic, glass, or metal trays containing many wells and depressions have been used for miniaturized microbiological studies (5–9). Commercially available divided petri dishes (10, 11) and sterilized plastic ice cube trays (12) have also been utilized. Microtiter plates containing 96 wells each were employed as miniaturized vessels for a number of tests (5, 13–15). Technical descriptions of the applications of the Microtiter system to many areas of microbiological work were compiled by Conrath (16).

Multiple-inoculation devices markedly increase the efficiency of culture transfer. Various types of apparatus have been used for such purposes; examples are bolts, needles, and pins of several kinds (17–28); plastic "stamps" (29, 30); velveteen (31–35); syringes (36, 37); Pasteur pipettes (38, 39); capillary tubes (40, 41); and a "piggy-back" method (42). Problems involved in replica plating by velveteen and plastic "stamp" methods are smearing of colonies and limitation to about 10 replications from one master plate. Clogging by agar may occur when Pasteur pipettes are used to inoculate solid media. Also, capillary tubes, Pasteur pipettes, and syringes can only be used on solid surfaces and may result in splattering on agar surfaces. Instruments containing pins and needles have the advantage of delivering test organisms to either liquid or solid media; splattering is minimal.

Provision must also be made for accurate observation of biochemical changes that result from microbial activities in these small culture chambers. Typical examples are the need for mass transfer of reagents into small-volume cultures in IMViC tests, trapping of gas in miniaturized carbohydrate fermentation tests, and reading of litmus milk reactions in small-volume tests.

This chapter describes several routine tests that have been miniaturized to affect rapid characterization of bacterial cultures. Bacterial isolates from turkey products were utilized to illustrate the procedures that were developed and the results that can be obtained by use of miniaturized microbiological techniques.

APPARATUS AND PROCEDURES

The general procedure for miniaturized microbiological techniques involves obtaining pure cultures, preparation of master plates, aseptic multiple inoculation into liquid and solid media, observation of biochemical changes, collection of data, and interpretation (Fig. 1). Not all tests were used for all cultures; only specific examples are cited herein.

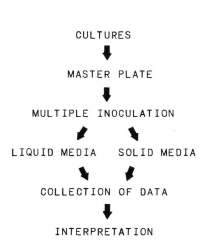

OPERATIONAL STEPS

CULTURES

MASTER PLATE

MULTIPLE INOCULATION

LIQUID MEDIA   SOLID MEDIA

COLLECTION OF DATA

INTERPRETATION

Fig. 1. Operation steps for miniaturized microbiological techniques.

Organisms

All named cultures were obtained from the culture collections of the Department of Food Technology, Iowa State University, Ames, and the Department of Microbiology, Pennsylvania State University, University Park. Identities of the named bacterial cultures were confirmed by using conventional bacteriological procedures, including morphology and gram reaction, cultural and biochemical characteristics, and (in some cases) serological reactions. The characteristics of the named bacteria corresponded well with their descriptions in *Bergey's Manual* (43). In addition to the named strains, 100 isolates that were obtained from turkey meat that was thought to be the vehicle of a staphylococcal food-poisoning outbreak, and 72 isolates from turkey rolls and roasts were used as test cultures for certain miniaturized techniques. The cultures were grown in trypticase soy broth (Difco) contained in test tubes or dropping bottles (Fig. 2; also see Ref. 40) at 37°C for 24 hours before they were utilized as inocula.

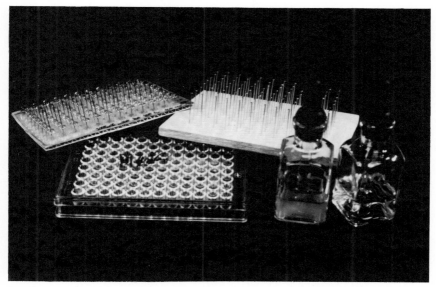

Fig. 2. Basic apparatus for miniaturized microbiological techniques: pinhead multiple-inoculator, pinpoint multiple-inoculator, Microtiter plate with plastic cover, dropping bottle (turbid) containing a culture, and dropping bottle containing sterile medium.

Media and Tests

Most of the miniaturized tests reported here were conducted in the wells of Microtiter plates. The Microtiter plates and covers were obtained from the Cooke Engineering Company, Alexandria, Virginia. As we described in earlier publications (12, 14), each Microtiter plate contains 96 wells; we routinely used 0.2 ml of medium per well. Sterile plate covers are available; these prevent contamination and retard evaporation during storage and incubation of the plates. We usually prepared the plates in the morning and inoculated them in the afternoon.

Microtiter plates of broth media were prepared by pipetting 0.2 ml of sterile medium into each well. The minimum quantity of medium that could be used was determined by adding different volumes of trypticase soy broth

(Difco) to the wells of a series of Microtiter plates. The plates were inoculated (see below) and were examined visually for the appearance of turbidity at various times following inoculation (Table 1). When inoculated with either *Staphylococcus aureus* or *Escherichia coli*, wells containing 0.05 to 0.2 ml of broth became definitely turbid after only 2 hours of incubation at 37°C. Volumes smaller than 0.2 ml became definitely turbid in less than 2 hours; but evaporation presented a problem when very small volumes were incubated for 20 hours or longer at 37°C. Incubation for at least 24 hours was thought to be necessary before a test could be considered to be truly negative; therefore, we used 0.2 ml of medium per well in most tests that utilized Microtiter plates. As Table 1 indicates, visible growth occurred much more rapidly in the wells of Microtiter plates than in larger volumes of media in test tubes.

Liquid media could be dispensed into Microtiter plates by using a 96-channel Automatic Pipette and a 12-channel Cooms Rinser and Cell Dispenser (Cooke Engineering Co.) or similar types of apparatus (44, 45).

Table 1 Turbidity Development in Different Cultural Volumes[a,b]

Culture Vessel	Volume (ml)	0	1	2	3	4	8	20
Wells of a	0.05	±	±	+	+ +	+ +	+ +	+ +
Microtiter	0.10	±	±	+	+ +	+ +	+ +	+ +
plate	0.15	±	±	+	+ +	+ +	+ +	+ +
	0.20	−	±	+	+	+ +	+ +	+ +
	0.25	−	±	±	+	+ +	+	+ +
	0.30	−	±	±	+	+	+	+ +
Test tube	0.50	−	−	−	±	+	+	+ +
	1.00	−	−	−	−	+	+	+ +
	5.00	−	−	−	−	±	+	+ +
	10.00	−	−	−	−	−	±	+ +

Time[c] (hours)

[a]Medium: trypticase soy broth (Difco) incubated at 37°C.
[b]Organisms tested: *Staphylococcus aureus* 137 and *Escherichia coli*, inoculated with a pinhead device; both cultures yielded similar results.
[c]Code: −, no visible turbidity; ±, slight turbidity; +, definite turbidity; + +, very turbid.

For preparing one or two plates of the same medium, however, a hand-operated pipette was more practical. A variety of commercially available automatic pipettors can be utilized for this purpose.

Liquid media in Microtiter plates were used for inocula and for carbohydrate fermentation, IMViC, and litmus milk tests. For the carbohydrate tests, a Pasteur pipette was used to overlay each well, after inoculation, with a layer of sterile Amojell–mineral oil mixture [one part of Amojell (American Oil Co.) and one part of mineral oil]. The overlay served to trap gas, thus permitting observation of gas production after incubation of the plates. As noted by Fougerat et al. (46), overlays tend to decrease oxygen permeability; but we did not notice any inhibition of growth by the Amojell overlay.

Reagents were added to wells for the methyl red, Voges-Proskauer, and indole tests by several different procedures. Since only a small amount of methyl red was necessary for the detection of acid production, the methyl red was added simultaneously to a set of cultures, after incubation, by use of the pinhead multipoint inoculator that is described in a later paragraph. Loops were used to add reagents for the Voges-Proskauer test, and reagents for the indole test were added with Pasteur pipettes.

A capillary-tube procedure was developed for determining certain litmus milk reactions. The colors of the surfaces and (by use of a mirror) bottoms of the culture wells were first recorded. Then short capillary tubes (1 mm O. D.) were placed into the wells (Fig. 5), the liquid in the wells was drawn into the tubes by capillary action, and the litmus reactions could be observed.

For tests on agar-containing media, the media were dispensed into petri plates. Tests that closely approached conventional bacterial methodology involved the use of staphylococcus medium no. 110 with egg yolk (Difco), mannitol salt agar (Difco), and tellurite–polymyxin–egg yolk agar (Difco). These media were poured into disposable, 15×150 mm petri dishes and dried overnight in an incubator at $37°C$ before inoculation.

Miniaturized triple sugar iron (TSI) agar (Difco) slants and sulfide indole motility (SIM) agar (Difco) and Sellers' medium (Difco) "deeps" were prepared in the wells of Microtiter plates. To make the mini-agar slants, 0.2 ml of melted medium was aseptically added to each well of a Microtiter plate that was positioned almost perpendicular to the bench top. To make the deeps, the plates remained flat when the wells were filled.

A semiquantitative catalase test was developed to evaluate the catalase activity of individual colonies on agar surfaces. A capillary tube containing a small amount of 3% hydrogen peroxide was touched to a colony, left in place for about 10 sec, and then examined to estimate the quantity of oxygen that had formed. When the amount of gas generated from an individual colony forced the H_2O_2 column to the tip of the capillary tube, a $+ + +$ activity was assigned. Values of $+ +$ and $+$ represented respectively, many and few bubbles observable in the H_2O_2 column; "0" was recorded if no bubbles were observed.

The dye-diffusion coagulase test of Fung and Kraft (46) was used to evaluate coagulase production; a score of 0 represents no coagulation, 1 represents minimal detectable coagulation, and 12 represents strong coagulation. Although we reported (47) that the method was semiquantiative, recent studies (Judith A. Weber, personal communication) indicate that this is true only at very low levels of coagulase and that the method is extremely sensitive.

Detection of enterotoxin production, both qualitatively (enterotoxins A, B, and/or C) and quantitatively ($\mu g/ml$), was performed by using the Microtiter hemagglutination inhibition (HAI) test of Morse and Mah (48) as modified by Fung (49). A medium containing 3% protein hydrolysate broth (Mead Johnson International, Evansville, Ind.), 3% NZ-amine NAK (Sheffield Chemical, Norwich, NY.), and vitamins (thiamin, niacin, and pantothenate, each at a level of 0.5 $\mu g/ml$), was used for growth and enterotoxin production by selected staphylococcal isolates.

Some other routine procedures, such as the gram stain, were also used; these are not elaborated on here. Parallel studies, using conventional procedures, were also made, permitting comparison of the results with the miniaturized methods.

Multiple-Inoculation Procedures

A master plate was prepared by placing four drops (about 0.2 ml) of a rapidly growing bacterial culture into each well of a presterilized Microtiter plate. Alternatively, a plate could be charged with sterile medium and then each well inoculated from a different colony by use of sterile toothpicks; or, individual colonies could be emulsified in sterile broth and then used as inocula (5). When all the inocula had been added to the master plate, it was covered with a sterile plastic cover (Fig. 2).

Multiple-inoculation devices were designed to transfer cultures from the master plates to solid or liquid media (Fig. 2). These devices, with either pinheads or pinpoints protruding, were made by inserting 27-mm-long, stainless steel pins into a wood, plastic (14), Lucite (12), or metal template. The pinhead device was suitable for introducing cultures to liquid media or for inoculating flat agar surfaces; the pinpoint device was used to stab miniaturized solid or semisolid media.

The inoculation procedures were usually performed under a hood that previously had been maintained at a low contamination level with ultraviolet irradiation. Sterilization of the inoculation devices was achieved by dipping the protruding portions of the pins into alcohol for about 20 sec, followed by flaming for about 1 sec, as described by Fung and Hartman (12). A sterile inoculation device was charged by lowering it slowly into a master culture plate. The results of preliminary tests indicated that each pinhead delivered about 0.0006 ml and each pinpoint delivered about 0.0002 ml. To complete the culture transfers, the charged device was used to inoculate solid, semisolid, or liquid media, as in Fig. 3 and 4. If an agar-containing medium was inoculated, the device could be returned directly to the master plate to be recharged; if liquid media were inoculated, the device was usually sterilized by flaming before it was recharged to eliminate carryover of chemicals from one liquid medium to another. The mini-plates were incubated in conventional incubators, alongside the conventional tests.

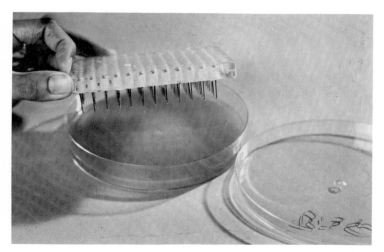

Fig. 3. Inoculation of 96 cultures onto an agar plate by the use of a multipoint inoculation device. The agar surface is dried overnight in a 37°C incubator prior to inoculation. After inoculation, the petri dish is covered and is incubated in an inverted position. Reproduced by permission of the National Research Council of Canada from *The Canadian Journal of Microbiology*, **18**, 1623–1627 (1972).

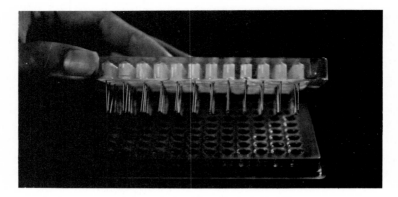

Fig. 4. Inoculation of 96 cultures into a liquid medium by use of a multipoint inoculation device. Reproduced by permission of the National Research Council of Canada from *The Canadian Journal of Microbiology*, **18**, 1623–1627 (1972).

RESULTS

The fermentation patterns of 25 bacteria on four media appear in Table 2. Although studies on carbohydrate fermentation patterns (acid production) in miniaturized vessels have been reported by others (5, 13), the gas-trapping procedure by Amojell is unique for this miniaturized system. Exact correlation was obtained between the Durham tube and Microtiter methods in the detection of acid production. Although the data are not given in Table 2, *Bacillus polymyxa* and *Salmonella thompson* produced only slight amounts of gas; gas was observed in only two or three of the four Durham tubes or Microtiter wells that were inoculated with each culture. These results indicate that the two methods produce approximately equivalent results, even if the test cultures yield only slight amounts of gas.

A comparison of miniaturized and conventional methods for IMViC tests on 24 gram-negative bacteria is presented in Table 3; perfect correlation was obtained. Tests were conducted to determine the optimal amounts of reagents to be transferred to the mini-cultures for the IMViC tests. To detect indole production, 2 drops of Kovac's reagent was transferred by a Pasteur pipette to each well of a Microtiter plate; if a red layer formed on top of the broth, the test was positive. For the methyl red test, four transfers of methyl red into the wells containing cultures were made by use of the pinhead multiple-inoculation device; pink or red color development in a well denoted a positive reaction. For the V-P test, one loopful (Microtiter loop, 0.05-ml capacity) of alpha-naphthol, followed by one loopful of KOH, were

Table 2 Carbohydrate Fermentation Patterns

Carbohydrates[a,b]

Organisms	Glucose Micro[b]	Glucose Durham	Fructose Micro	Fructose Durham	Lactose Micro	Lactose Durham	Sucrose Micro	Sucrose Durham
Achromobacter parvulus	+	+	+	+	−	−	−	−
Agrobacterium radiobacter	−	−	−	−	−	−	−	−
Alcaligenes faecalis	−	−	−	−	−	−	−	−
Arthrobacter globiformis	−	−	−	−	−	−	−	−
Bacillus polymyxa	+	+	⊕	⊕	+	+	+	+
Bacillus megaterium	+	+	+	+	−	−	+	+
Bacillus sulfidus	+	+	+	+	−	−	−	−
Corynebacterium xerosis	+	+	+	+	+	+	+	+
Enterobacter aerogenes	⊕	⊕	⊕	⊕	⊕	⊕	⊕	⊕
Enterobacter cloacae	⊕	⊕	⊕	⊕	⊕	⊕	⊕	⊕
Escherichia coli	⊕	⊕	⊕	⊕	⊕	⊕	−	−
Gaffkya tetragena	+	+	+	+	+	+	+	+
Micrococcus rhodochrous	+	+	−	−	−	−	−	−
Mycobacterium phlei	+	+	−	−	−	−	−	−
Proteus vulgaris	+	+	+	+	−	−	+	+
Pseudomonas aeruginosa	−	−	−	−	−	−	−	−
Salmonella typhimurium	⊕	⊕	⊕	⊕	−	−	−	−
Salmonella typhosa	+	+	+	+	−	−	−	−
Salmonella paratyphi	⊕	⊕	⊕	⊕	−	−	−	−
Salmonella pullorum	⊕	⊕	⊕	⊕	−	−	−	−
Salmonella thompson	⊕	⊕	⊕	⊕	−	−	−	−
Sarcina lutea	−	−	−	−	−	−	−	−
Shigella flexneri	+	+	+	+	−	−	−	−
Staphylococcus aureus	+	+	+	+	+	+	+	+
Streptococcus durans	+	+	+	+	+	+	−	−

[a] Other carbohydrates tested were dulcitol, galactose, glycerol, inulin, maltose, mannitol, raffinose, and sorbitol.

[b] Code: Micro, Microtiter-Amojell overlay technique; Durham, Durham inverted tube technique.

[c] +, acid production (yellow); −, no reaction (red); ⊕, acid and gas production.

Table 3 Comparison of Miniaturized (Mini) and Conventional (Conv) Methods for IMViC Tests[a]

Test Organisms	Indole Mini 8–12 hours	Indole Conv 24 hours	M–R Mini 48 hours	M–R Conv 120 hours	V–P Mini 24 hours	V–P Conv 48 hours	Citrate Mini 8–12 hours	Citrate Conv 24 hours
Achrombacter parvulus	−	−	−	−	+	+	−	−
Agrobacterium radiobacter	−	−	−	−	−	−	−	−
Alcaligenes faecalis	−	−	−	−	−	−	+	+
Corynebacterium xerosis	−	−	−	−	−	−	−	−
Enterobacter aerogenes 11	−	−	−	−	+	+	+	+
Enterobacter aerogenes 11a	−	−	−	−	+	+	+	+
Enterobacter cloacae	−	−	−	−	+	+	+	+
Escherichia coli A-1	+	+	+	+	−	−	−	−
Escherichia coli B	+	+	+	+	−	−	−	−
Escherichia coli C-30	−	−	+	+	−	−	−	−
Proteus vulgaris 9434	−	−	+	+	−	−	−	−
Proteus vulgaris X19	−	−	+	+	−	−	−	−
Pseudomonas aeruginosa	−	−	−	−	−	−	+	+
Salmonella paratyphi	−	−	+	+	−	−	+	+
Salmonella pullorum	−	−	+	+	−	−	−	−
Salmonella thompson	+	+	+	+	−	−	+	+
Salmonella typhimurium	−	−	+	+	−	−	+	+
Salmonella typhosa	−	−	+	+	−	−	−	−
Serratia marcescens	−	−	−	−	−	−	+	+
Shigella alkalescens	+	+	−	−	−	−	−	−
Shigella flexneri V	−	−	+	+	−	−	−	−
Shigella flexneri W	+	+	−	−	−	−	−	−
Shigella flexneri 9748	+	+	−	−	−	−	−	−
Shigella sonnei	−	−	+	+	−	−	−	−

[a]The indoles, M–R, V–P, and citrate tests were made as described in the text. The times given are the minimum incubation periods necessary for all positive cultures to respond to the miniaturized test or the recommended incubation period for the conventional test; +, positive test; −, negative test.

Reproduced by permission of the International Association of Milk, Food, and Environmental Sanitarians. These data were reported in Reference 15.

added to each culture; this combination yielded the best results of various combinations that were examined to detect the presence of acetyl-methyl-carbinol; a positive test was the formation of a pink or red color in a well.

Table 4 shows the biochemical reactions obtained on 23 test organisms in mini-TSI agar slants and conventional TSI agar slants. Although there was a perfect correlation of data between the two methods in 18 species, four cultures (*Agrobacterium radiobacter*, *Bacillus polymyxa*, *Gaffkya tetragena*, and *Micrococcus rhodochrous*) gave weak–positive reactions in the mini-slants and negative reactions in the conventional tubes. This may be the result of a high ratio of inoculum size versus volume of medium in the mini-slants. It is

Table 4 Comparison of Miniaturized and Conventional TSI Test

Test Organism	Microtiter[a] Butt	Slant	Time[b] (hours)	Conventional[a] Butt	Slant	Time[b] (hours)
Achromobacter parvulus	Ac	Al	10–20	Ac	Al	20
Agrobacterium radiobacter	Ac	Al	10–20	NC	NC	36
Alcaligenes faecalis	Al	Al	5–10	Al	Al	20
Bacillus megaterium	Ac	Ac	5–10	Ac	Ac	20
Bacillus polymyxa	Ac	Al	10–20	NC	NC	36
Bacillus sulfidus	Ac	Al	10–20	Ac	Al	36
Corynebacterium xerosis	Ac	Al	10–20	Ac	Al	20
Enterobacter aerogenes 11	Ac	Ac	5–10	Ac	Ac	20
Enterobacter cloacae	Ac	Ac	5–10	Ac	Ac	20
Escherichia coli	Ac	Ac	5–10	Ac	Ac	20
Gaffkya tetragena	Ac	Ac	10–20	NC	NC	36
Micrococcus rhodochrous	Ac	Ac	10–20	NC	NC	36
Mycobacterium phlei	Ac	Ac	10–20	Ac	Ac	20
Proteus vulgaris	Ac	Ac	10–20	Ac	Ac	20
Pseudomonas aeruginosa	Al	Al	5–10	Al	Al	20
Salmonella typhimurium	Ac	Al, H_2S	10–20	Ac	Al, H_2S	20
Salmonella typhosa	Ac	Al	10–20	Ac	Al, H_2S	20
Salmonella pullorum	Ac	Al, H_2S	10–20	Ac	Al, H_2S	20
Salmonella thompson	Ac	Al, H_2S	10–20	Ac	Al, H_2S	20
Sarcina lutea	NC	NC	10–20	NC	NC	36
Shigella flexneri	Ac	Al	10–20	Ac	Al	20
Staphylococcus aureus	Ac	Ac	10–20	Ac	Ac	20
Streptococcus faecalis	Ac	Ac	5–10	Ac	Ac	20

[a]Ac, acid; Al, alkaline; NC, no change; H_2S, hydrogen sulfide.
[b]Time for completion of reaction.

also possible that the degree of oxygen availability is influenced by the type, size, and shape of container; the Microtiter slants might be more highly oxygenated than the conventional slants, and oxidative reactions might be accentuated under such conditions. *Salmonella typhosa* showed slight hydrogen sulfide production in the conventional test but not in the mini-slants, which probably means that the mini-slants are not as sensitive as the conventional method in detection of H_2S production. This is to be expected because although H_2S is water-soluble, it is also volatile and could escape readily under the assay conditions. An Amojell overlay or deeper wells might correct this deficiency of the mini-slant TSI test. The time for completion of the TSI-agar reactions varied among species in the mini-slant tests. Some species produced "final" reactions after only 5 to 10 hours of incubation, but 16 species required between 10 and 20 hours of incubation to produce "final" reactions. In some cases, prolonged incubation of the mini-slants resulted in changes of color from acid to alkaline; this problem may be corrected by using multiple-well plates containing deeper wells. Miniaturization of the TSI-agar test requires more study.

Figure 5 shows the results of some litmus milk reactions in miniaturized culture plates using the capillary tube procedure. No medium was drawn

Fig. 5. Capillary tube detection of litmus milk reactions. No medium was drawn into the fifth tube from the left, indicating that the milk in that well was coagulated. Color of the milk is readily observable once the milk is drawn into the capillary tube.

into the fifth tube from the left, indicating that the milk in that well was coagulated. The sixth tube from the left showed a small milk clot at the bottom; whey was drawn into the capillary tube. The color of the litmus milk was readily observable when the milk was drawn into the capillary tube. A 24-hour incubation period was best for completion of mini-litmus milk tests, although most reactions occurred after only 8 hours of incubation.

Examples of miniaturized SIM tests are presented in Fig. 6. Note that the pinpoint device was a very efficient means for stabbing these agar deeps. Cultures A-1 and A-2 showed growth without motility and H_2S production. Culture B-3 showed growth and H_2S production and no motility. Cultures A-3 and B-2 showed growth and motility but no H_2S production. Culture B-1 showed growth with motility and H_2S production. Eight hours of incubation was sufficient for detection of growth and motility in mini-SIM cultures, but development of H_2S may not be apparent until after 12 hours of incubation.

Sellers' medium was designed to differentiate nonfermentive gram-negative bacteria from fermentative bacteria. Figure 7 indicates that *Alcali-*

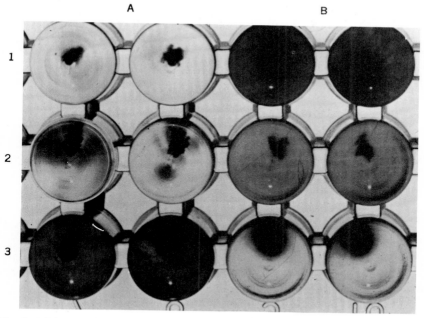

Fig. 6. Miniaturized SIM tubes. All cultures, presented in duplicate, grew. Cultures A-1 and A-2: motility−,H_2S−; A-3 and B-2: motility+,H_2S−; B-1: motility+,H_2S+; B-3: motility−,H_2S+.

genes faecalis and *Pseudomonas aeruginosa* (the four wells with darker color) produced typical reactions by the miniaturized Sellers' procedure. Eight hours of incubation was sufficient for completion of reactions in this test.

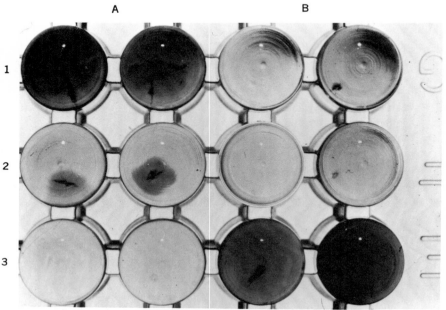

Fig. 7. Miniaturized Sellers' medium. Cultures were tested in duplicate. Culture A-1 (*Pseudomonas aeruginosa*): green; A-2 to B-2 (four different fermentative bacterial): yellow; B-3 (*Alcaligenes faecalis*): blue.

Using capillary tubes containing 3% H_2O_2, a semiquantitative catalase activity scale was developed (Fig. 8). Table 5 compares capillary-tube and conventional techniques for the detection of relative catalase activities of 18 cultures grown on nutrient agar and blood agar. We note with particular interest that for the conventional method all blood agar plates exhibited large amounts of gas when a drop of 3% H_2O_2 was applied, regardless of the type of organism growing on the plate; since blood contains catalase, H_2O_2 can be broken down even in the absence of bacterial catalase. On the contrary, the capillary-tube method could distinguish catalase-positive from catalase-negative organisms on blood agar plates. Also, catalase activities of isolated colonies in the midst of mixed cultures (data not shown) could be detected by the capillary-tube technique but not by the conventional technique in both nutrient agar and blood agar. Colony size had no direct relation to the numbers of bubbles obtained in capillary tube tests. This capillary catalase test merits further evaluation for use in clinical diagnostic microbiology.

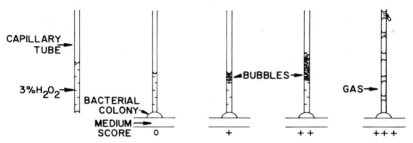

Fig. 8. Capillary tube procedure for semiquantitative catalase test. Positive reactions were usually instantaneous after contact of the H_2O_2 with the colonies; 10 sec was used as an arbitrary time limit before reading rhe results of this test.

Utilizing the miniaturized replica plating techniques and other tests, 100 isolates from turkey meat were characterized (Table 6, notes *a* and *b*). Ninety-seven isolates were *Staphylococcus aureus*; 37 were enterotoxigenic (Table 6). In another study, 72 isolates from turkey rolls and roasts were characterized (Table 7, notes *a* and *b*). Sixty isolates were *S. aureus* and 12 were *S. epidermidis*; five stock cultures also were examined. Twenty-seven of the 60 isolates of *S. aureus* were enterotoxigenic.

DISCUSSION

We have presented a series of ideas that can be incorporated into convenient systems for the rapid characterization of bacteria. Although neither the mass inoculation procedure nor small-volume tests are new concepts, we have described several unique features that should facilitate characterization of bacteria in certain miniaturized tests. Examples are: the Amojell overlay technique for trapping gas in mini-fermentation tests, mass transfer of reagents into growth cultures in the mini-methyl red test, detection of motility and biochemical tests in mini-semisolid media and mini-agar slants, the use of capillary tubes for semiquantitative detection of catalase activities of pure and mixed cultures, and observation of litmus milk reactions by a capillary-tube procedure.

Table 5 Capillary Tube and conventional Techniques for Semiquantitative Measurement of Catalese Activity

Test Organisms	Nutrient Agar		Blood Agar[a]	
	Mini[b]	Conv[c]	Mini[b]	Conv[c]
Bacillus cereus	+ + +	L	+ + +	L
Bacillus megaterium	+ + +	L	+ + +	L
Enterobacter aerogenes	+ +	S	+	L
Escherichia coli	+ +	S	+ +	L
Gaffkya tetragena	+	S	+ +	L
Micrococcus varians	+ +	L	+ + +	L
Pseudomonas aeruginosa	+ + +	L	+ + +	L
Sarcina lutea	+ + +	L	+ + +	L
Staphylococcus aureus 241a	+ + +	L	+ + +	L
Staphylococcus aureus 241c	+ + +	L	+ + +	L
Staphylococcus aureus 241d	+ + +	L	+ + +	L
Staphylococcus aureus 244	+ + +	L	+ + +	L
Streptococcus faecalis	0	0	0	L
Streptococcus faecalis 249	0	0	0	L
Streptococcus faecalis 249a	0	0	0	L
Streptococcus faecalis 250	0	0	0	L
Streptococcus faecalis 251	0	0	0	L
Streptococcus faecalis 254	0	0	0	L

[a]When a drop of 3% H_2O_2 was placed on the blood agar medium without colonies, good effervescence was observed. When a capillary tube containing 3% H_2O_2 was stabbed into the blood agar, a + + score was obtained. If the capillary tube touched the surface of the agar lightly, however, no gas bubbles were generated.

[b]Mini, miniaturized method; capillary tube contacted with bacterial colonies; 0, +, + +, and + + + represent no bubble, a few bubbles, many bubbles, and gas forcing the 3% H_2O_2 to the tip of the capillary tube, respectively. Data were obtained from 5 or more individual colonies for each determination.

[c]Conv, conventional method; one drop of 3% H_2O_2 on the colony; 0, S, and L represent no bubbles, small numbers of bubbles, and large numbers of bubbles, respectively.

Table 6 Enterotoxin Production and Other Characteristics[a] of 100 Isolates[b] from Turkey Meat

Organisms,[c] Number of Isolates	Enterotoxin A[d] (μg/ml)	Coagulase Score
1	1.77	11
1	1.77	10
2	1.77	9
1	1.77	8
1	1.77	7
2	1.77	6
3	1.77	5
1	1.77	4
1	1.18	11
3	1.18	10
1	1.18	4
1	0.89	11
2	0.89	10
2	0.89	9
1	0.89	8
2	0.89	7
2	0.89	5
1	0.89	4
4	0.89	3
1	0.59	12
2	0.59	11
2	0.59	10
60	Negative	2–12
3	Negative	Negative

[a]All isolates were gram-positive cocci except three isolates, which were gram-positive rods. All isolates grew on nutrient agar and staphylococcus medium 110 with egg yolk. All isolates except six grew on tellurite–polymyxin–egg yolk agar.

[b]All isolates were obtained from typical colonies on staphylococcus medium 110 with egg yolk.

[c]The isolates were arranged according to the amount of toxin produced and then were subordered according to coagulase score.

[d]No enterotoxins B or C detected.

Reproduced by permission of the National Research Council of Canada from *The Canadian Journal of Microbiology*, **18**, 1623–1627 (1972).

Table 7 Enterotoxin Production and Other Characteristics of 72 Isolates and 5 Named Cultures[a]

Organisms	Enterotoxin (μg/ml)			Coagulase Score	Carbohydrate Fermentation Patterns[b]
	A	B	C		
8 Isolates	0.89	—	—	5–9	A
5 Isolates	0.59	—	—	5–9	
5 Isolates	—	—	0.89	5–7	
2 Isolates	—	—	0.59	6, 7	
2 Isolates	0.89	—	0.89	5, 11	
1 Isolate	0.89	—	0.59	6	
2 Isolates	0.59	—	0.89	5, 12	
2 Isolates	0.59	—	0.59	4, 6	
31 Isolates	—	—	—	4–12	B
2 Isolates	—	—	—	7	C
12 Isolates	—	—	—	Negative	D
S. aureus 100	—	—	—	11	A
S. aureus 196E	2.36	—	—	12	
S. aureus S-6	2.36	220.00	—	12	
S. aureus 137	—	—	7.02	10	
S. aureus 494	—	—	—	11	

[a]All isolates were obtained from typical colonies on staphylococcus medium 110 with egg yolk. All isolates and known *S. aureus* were gram-positive cocci.

[b]Carbohydrate fermentation patterns: A is positive for glucose, lactose, maltose, mannitol, sorbitol, sucrose, and xylose; negative for dulcitol and salicin. B is same as A except one isolate was negative for maltose. C is same for A, except both were negative for mannitol. D is positive for glucose (8/12), lactose, maltose (8/12), and sucrose (8/12); negative for dulcitol, mannitol, salicin, sorbitol (8/12), and xylose (8/12).

We have also presented data showing the usefulness of some of these miniaturized techniques in the characterization of staphylococci. Using some of the procedures described in this paper, Fung and Miller (50) studied the effects of 42 dyes on the growth of 30 bacterial cultures; Wood (51) characterized the fermentation patterns of 33 *Bacillus* spp. on ethanol, glycerol, and various sugar and sugar alcohols; and Barrile et al. (52)

obtained carbohydrate fermentation patterns of 277 isolates obtained from cocoa beans with great efficiency. J. A. Davies and coworkers (13, 53) also have used Microtiter plates for rapid carbohydrate fermentation tests.

Many other microbiological tests can be miniaturized, depending on the needs of the individual investigator, and a number of these tests can be automated. Analyses of the large amounts of data generated by these miniaturized tests could be facilitated by computers.

We estimate that these techniques require about 5% of the materials and 10% of the time needed for conventional procedures in studies involving many cultures. Savings of laboratory and incubator space and reductions of incubation times are further advantages of the miniaturized tests. Furthermore, these principles are readily adaptable to other laboratory procedures.

Acknowledgments

The senior author wishes to extend his appreciation to the Dynatech Corporation (Cooke Engineering Company, Alexandria, Va., Dynatech Ag, Zug, Switzerland) and to T. B. Conrath and A. C. Thorne, for making possible his trip to the symposium in the form of a traveling grant. The discussions and assistance of C. L. Fung, A. A. Kraft, R. D. Miller, K. A. Culkin, L. L. VandenBosch, D. T. Petrishko, and M. Schmoyer are greatly appreciated.

REFERENCES

1. P. A. Hartman, *Miniaturized Microbiological Methods*, Academic Press, New York, 1968.
2. B. M. Gibbs and F. A. Skinner, Eds., *Soc. Appl. Bacteriol. Tech., Ser. 1*, 145 pp, 1966.
3. B. M. Gibbs and D. A. Shapton, Eds., *Soc. Appl. Bacteriol. Tech., Ser. 2*, 212 pp, 1968.
4. A. Baille and R. J. Gilbert, Eds., *Soc. Appl. Bacteriol. Tech., Ser. 4*, 233 pp, 1970.
5. C. N. Huhtanen, J. Naghski, and E. S. Dellamonica, *Appl. Microbiol.*, **24**, 618 (1972).
6. R. A. Beargie, E. C. Bracken, and H. D. Riley, Jr., *Appl. Microbiol.*, **13**, 279 (1965).
7. W. F. Hink and J. D. Briggs, *J. Insect Physiol.*, **14**, 1025 (1968).
8. M. Kende and M. L. Robbins, *Appl. Microbiol.*, **13**, 1026 (1965).
9. G. Takatsy, *Acta Microbiol. Acad. Sci. Hung.*, **3**, 191 (1955).
10. P. H. A. Sneath and M. Stevens, *J. Appl. Bacteriol.*, **30**, 495 (1967).
11. I. R. Hill, *Soc. Appl. Bacteriol. Tech. Ser.*, **4**, 175 (1970).
12. D. Y. C. Fung and P. A. Hartman, *Can. J. Microbiol.*, **18**, 1623 (1972).
13. J. A. Davis, J. R. Mitzel, and W. E. Beam, *Appl. Microbiol.*, **21**, 1072 (1971).
14. D. Y. C. Fung and R. D. Miller, *Appl. Microbiol.*, **20**, 527 (1970).
15. D. Y. C. Fung and R. D. Miller, *J. Milk Food Technol.*, **35**, 328 (1972).

16. T. B. Conrath, *Handbook of Microtiter Procedures*, Dynatech Corp., Cambridge, Mass. 1972.

17. W. B. Cooke, *Mycopathol. Mycol. Appl.*, **25**, 195 (1965).

18. S. D. Garrett, *Brit. Mycol. Soc. Trans.*, **29**, 171 (1946).

19. L. J. Hale and G. W. Inkley, *Lab. Pract.*, **14**, 452 (1965).

20. R. U. Haque and J. N. Baldwin, *J. Bacteriol.*, **88**, 1442 (1964).

21. R. Holliday, *Nature (London)*, **178**, 987 (1956).

22. B. Lighthart, *Appl. Microbiol.*, **16**, 1797 (1968).

23. T. E. Lovelace and R. R. Colwell, *Appl. Microbiol.*, **16**, 944 (1968).

24. R. A. Murphy and R. U. Haque, *Amer. J. Clin. Pathol.*, **47**, 554 (1967).

25. W. A. Smirnoff and J. M. Perron, *J. Invertebr. Pathol.*, **7**, 320 (1965).

26. P. R. Watt, L. Jeffries, and S. A. Price, *Soc. Appl. Bacteriol. Tech. Ser.*, **1**, 125 (1966).

27. T. H. Wood and S. K. Mahajan, *Genet. Res.*, **15**, 335 (1970).

28. S. Zamenhof, *J. Bacteriol.*, **18**, 111 (1961).

29. T. Bergan, *Acta Pathol. Microbiol. Scand.*, **72**, 396 (1968).

30. H. D. Tresner and J. A. Hayes, *Appl. Microbiol.*, **19**, 186 (1970).

31. J. Lederberg and E. M. Lederberg, *J. Bacteriol.*, **63**, 399 (1952).

32. J. S. Lee and G. C. Wolfe, *Food Technol.*, **21**, 35 (1967).

33. M. A. Shiflett, J. S. Lee, and R. O. Sinnhuber, *Appl. Microbiol.*, **14**, 411 (1966).

34. M. Shifrine, H. J. Phaff, and A. L. Demain, *J. Bacteriol.*, **68**, 28 (1954).

35. P. F. Wiseman and W. B. Sarles, *J. Bacteriol.*, **71**, 480 (1956).

36. G. D. Anagnostopoulos, *J. Gen. Microbiol.*, **64**, 251 (1971).

37. J. J. Farmer, *Appl. Microbiol.*, **20**, 517 (1970).

38. R. L. Massey and R. H. T. Mattoni, *Appl. Microbiol.*, **13**, 798 (1965).

39. J. P. Seman, Jr., *Appl. Microbiol.*, **15**, 1514 (1967).

40. P. A. Hartman and P. A. Pattee, *Appl. Microbiol.*, **16**, 151 (1968).

41. M. Valland, M. Sc. thesis, Iowa State University, Ames, Iowa, 1969.

42. L. W. Catalano, Jr., D. A. Fuccilo, and J. L. Sever, *Appl. Microbiol.*, **18**, 1094 (1969).

43. R. S. Breed, E. G. D. Murray, and N. R. Smith, *Bergey's Manual of Determinative Bacteriology*, William and Wilkins, Baltimore, 1957.

44. P. A. M. Guinee, W. J. Van Leeuwen, and W. H. Jansen, *Appl. Microbiol.*, **23**, 1172 (1972).

45. G. Middlebrook, Z. Reggiardo, and G. R. Taylor, *Appl. Microbiol.*, **20**, 852 (1970).

46. J. Fougerat, J. Buissiere, and J. Lahneche, *Rev. Inst. Pasteur Lyon*, **4**, 215 (1971).

47. D. Y. C. Fung and A. A. Kraft, *Appl. Microbiol.*, **16**, 1608 (1968).

48. S. A. Morse and R. A. Mah, *Appl. Microbiol.*, **15**, 58 (1967).

49. D. Y. C. Fung, Ph. D. dissertation, Iowa State University, Ames, Iowa, 1969.

50. D. Y. C. Fung and R. D. Miller, *Appl. Microbiol.*, **25**, 793 (1973).

51. R. T. Wood, Ph. D. dissertation, Pennsylvania State University, University Park, Pa., 1970.

52. J. C. Barrile, K. Ostovar, and P. G. Keeney, *J. Milk Food Technol.*, **34**, 369 (1971).

53. J. A. Davies and R. R. Gutekunst, *Bacteriol. Proc.*, **1968**, 84 (1968).

CHAPTER 22

Development of Reagent-Impregnated Test Strips* for Identification of Microorganisms

DONALD P. KRONISH

*PathoTec® General Diagnostics Division, Warner-Lambert Co. Morris Plains, New Jersey 07950 USA

INTRODUCTION

In developing any new approach to a technology or in improving an existing one, the first step always should be a critical examination of the current state of the art, including a definition of areas that are deficient or undesirable. Only at this point can a rational optimum be proposed. The optimum thus indicated may not be achievable in a practical sense with available technology, or at a reasonable cost, but it can be defined.

The most desirable solution to many of the problems associated with the processing of a sample in a clinical microbiology laboratory would consist of one small black box, having a reasonable cost, into which samples from any source could be inoculated. The operator would merely push a button and this presently undefined machine would provide an immediate printout containing the name of the offending organism or organisms, antibiotic susceptibility patterns, and the epidemiological significance of all organisms in the specimen in relation to similar isolates in the hospital, the city, or the geographical area. Such an optimal system may be available in the future, but for the present, advances in microbiological technology are less ambitious.

What is the current applied state of the art, and why for the most part, has the clinical microbiology laboratory not benefited significantly from the major technological advances occurring elsewhere in the clinical laboratory? Let us also examine areas in which meaningful changes can be made in both technological and economic terms.

There are clearly three principal areas of responsibility in the clinical microbiology laboratory. The method of handling individual samples varies substantially; but after receiving a proper specimen that has been properly taken and properly handled, the laboratory must isolate microorganisms that may be responsible for disease. It must identify these organisms and, finally, it must provide antibiotic susceptibility data for therapy.

PROCEDURAL BACKGROUND

Let us consider some of the parameters of a typical procedure for handling a sample. I do this with some trepidation, since few things are as atypical as a typical procedure in a microbiology laboratory. However, an examination of sample handling procedures can serve as a useful measure of the current state of the art, and it will point up areas for the potential application of new technology.

For the purposes of discussion, assume that a valid specimen arrives in the laboratory on a Monday morning. It is plated onto two or more selective or differential primary isolation media (after enrichment, if necessary) and incubated under the proper conditions for optimum growth. On Tuesday morning, the plates are examined, and several suspect colonies of each morphological type are inoculated to sugar differential media (TSI, Russell's, Kligler's, etc.) and a Kirby-Bauer susceptibility test is set up with 5 colonies from the primary plate. Sugar differential media are incubated until Wednesday morning, at which time an appropriate set of biochemical tests is arranged to provide an identification. On Thursday morning, most of the biochemicals can be interpreted and an identification made, but some biochemicals must be incubated for periods of up to 4 days before a definitive negative can be read.

This is obviously not a satisfactory procedure. The value of the answer generated in the laboratory is a direct function of the speed with which data are provided to the physician. In many areas the laboratory is constantly under pressure for an answer, but an *accurate* answer cannot be provided rapidly. As a result of this constant requirement for speed, the laboratory feels forced to make potentially dangerous judgments directly from the primary plate, using characteristics of colony morphology and appearance. The generation of more complete data is time-consuming because each time an organism is inoculated to a different medium, it must grow before a physiological or biochemical event takes place and before a judgment can be made.

SYSTEM CHARACTERISTICS

A bacteriological medium containing both growth support and selective and/or inhibitory characteristics is obviously essential for the isolation of an organism from a specimen–but one of the questions raised by a so-called typical isolation and identification procedure is, Why is it necessary to grow and often regrow an isolated organism several times under somewhat unfavorable conditions (i.e., growth support systems take critical time, and growth support systems, by definition, have limited stability characteristics and limited reproducibility)? The only reasonable answer is that this is the present state of the art. The state of an art, however, is not necessarily a measure of the most desirable characteristics of a given technology. What are some of the basic characteristics of all conventional bacteriological systems which are undesirable in terms of the job they are required to do?

1. Conventional systems do not generate data rapidly because each has a dual purpose. Before a biochemical event can take place, organisms must

proliferate to numbers which will metabolize a given substrate to a specific end product. In addition, conventional systems typically contain relatively large amounts of material to provide maximum stability.

2. Conventional media systems are inherently variable. Almost all contain complex peptones, carbohydrates, and accessory growth factors that are not or cannot be defined from batch to batch or from year to year. Since the judgment made from a given medium should be the result of an objective measurement, the components of that measurement should be as constant as possible.

3. Conventional media are unstable—almost by definition. Changes in the aqueous/solids ratio can and do affect the quality of a judgment, and frequently, components of a complex media system react with each other after preparation.

4. Since all conventional media systems are designed to support growth of microorganisms, these media will support growth whether inoculation is intentional or accidental. A single unnoticed colony on an agar medium can completely invalidate the measurement of a biochemical parameter.

5. When an end product of metabolism is being measured by other than pH changes, it is almost always necessary to add reagents to the system. The addition of acidified PDAB for the indole test, naphthylamine and sulfanilic acid for the nitrate reductase test, or $FeCl_3$ required for the PD test are, at best, inconvenient, but—

6. Many of these reagents are also unstable and require special conditions for proper storage.

7. Finally, as a sum of all these characteristics, an objective interpretation of a chemical reaction that takes place in a thoroughly undefined and unstandardized system frequently cannot be as objective as is required.

However, two basic concepts of conventional media systems should be maintained in the development of any new approach to technology. These are, first, flexibility in design of identification procedures to fit a laboratory routine in terms of information available from the specimen and primary media. Second, test procedures should be maintained as discrete systems that measure a single physiological or biochemical event. Measurement of multiple parameters within a single frame can lead to misinterpretations and inaccurate data.

STRIP TECHNOLOGY

Within this outline of disadvantages and advantages of conventional media systems, several new approaches have been developed for the biochemical characterization of microorganisms.

A strip of specially purified filter paper having certain arbitrary dimensions, and used as an inert support, can be impregnated with specific quantities of specific reagents, forming a simple test system. A simple example is the application of reagents for detection of the enzyme cytochrome oxidase. Several of the key concepts of reagent-impregnated strip technology are demonstrated in this test system. Reagents can be applied to paper with high levels of precision ensuring reproducibility. Following reagent application, strips are dried to a moisture specification of less than 3% and maintained at that level by a dessicant. Oxidase reagents are unstable when mixed in solution, but they are stable when dry and maintained under proper reducing conditions. Therefore, "dry" means "stable" in this and other test systems.

In measuring most biochemical reactions of microorganisms, there is a need to provide intimate contact between bacteria and substrate for a period of time before a reaction is measurable. Absorbant paper would appear to be unsatisfactory, since a restricted contact area is required. Therefore, a waterproof barrier is applied to the paper to prevent chromatographic migration. This barrier is a combination of an acrylic polymer and a chrome complex polymer.

After localization of the area on the paper in which a reaction may take place, it is possible to apply specific substrates and a pH indicator to the strip and to add a color identification band to the distal end. Since time is one of the real differences between strip technology and conventional media, the amount of material applied to paper is important in addition to the nature of the substrate. Since it takes less time for an organism to metabolize a small amount of substrate than a larger amount of substrate, micro amounts of substrate plus indicator are applied to the strip.

Substrate and detection reagents can be combined in a single reagent zone only when there is no metabolic interference between the two materials and no chemical incompatibility. This is the case with urease and lysine decarboxylase test systems.

There are, however, several biochemical reactions in which reagents required to detect an end product of metabolism are toxic to, or interfere with the production of, the end product. Such reactions include nitrate reductase, phenylalanine deaminase, indole, and Voges-Proskauer reactions. Under these conditions, detection reagents must be separated from substrate. Since the design flexibility of an inert paper support allows impregnation of reagents at any point, such reagents as acidified PDAB, naphthylamine and sulfanilic acid, or a ferric salt, are applied to the paper in an area separated from substrate by a waterproof barrier. A second waterproof barrier is then applied to the strip to localize color formation.

In addition, reagents in some biochemical test systems are incompatible with each other. These reagents may be in substrate or detection systems. Equipment has been developed to apply one reagent to the surface of the inert paper support such that it is absorbed halfway through the thickness of the paper, and another reagent to the other side, such that the reagents meet in the center of the paper but do not mix. Such is the case with both the malonate utilization test strip and the detection zone of the nitrate reductase test strip. In the nitrate reductase test strip both naphthylamine and sulfanilic acid are most stable at an alkaline pH of about 11. They are not stable at an acid pH. However, an acid pH is necessary for the diazotization of sulfanilic acid, prior to its reaction with naphthylamine. Therefore, sulfanilic acid and naphthylamine are applied to one side of the filter paper at an alkaline pH and a crystalline acid is applied to the other side. The result in both cases is prolonged stability.

The physical characteristics of paper as an inert support for reagents are highly versatile. Since all *Enterobacteriaceae* produce at least some hydrogen sulfide, the taxonomic value of this parameter would be negligible except that in TSI agar only some enterics are H_2S positive. The selectivity of a positive reaction in this medium is due to the detection system: ferrous sulfate and sodium thiosulfate. Under proper reduction-oxidation conditions, ferrous sulfate is oxidized to the ferric salt and forms a black precipitate with H_2S that has been produced by an organism. The time required for an organism to achieve the proper oxidation-reduction conditions in TSI agar makes it necessary to examine this medium for H_2S production after both 24 and 48 hours. After 48 hours care should be exercised in interpretation of weak and delayed reactions.

In a reagent-impregnated strip system, speed is an important characteristic, and delayed reactions are not useful. Therefore, H_2S formed from substrates on the strip by an *E. coli* is absorbed quantitatively on the strip by a carbonate zone and excess H_2S produced by TSI-positive cultures react with lead acetate impregnated on the strip above the carbonate. Therefore, data normally generated in 24 or 48 hours can be measured in 4 hours.

Strip technology is also highly versatile, and paper can also be used as an inert support for reagents in a sequential chemical reaction for measuring niacin produced by mycobacteria. Some of the original work on this configuration was done by James Kilburn and George Kubica at the CDC.

The strip is impregnated with precise quantities of a primary aromatic amine, potassium thiocyanate, a crystalline acid, and a component known as chloramine T. As in all strip technology, the chemistry is conventional; after extraction of niacin from the mature cell mass, an aliquot is placed in a tube, the strip is added, and the tube is sealed. The niacin-containing extract mixes with the primary aromatic amine and the thiocyanate and the crystalline acid and the chloramine T. Concurrently, the addified thiocyanate reacts with the chloramine T to form a cyanogen halide, which ruptures the niacin ring structure to form the corresponding glutaconic aldehyde. The glutaconic aldehyde couples with the aromatic amine to form the corresponding Schiff base, which happens to be yellow.

Obviously, the niacin reaction has no utility among the *Enterobacteriaceae*. I am using it here only to further demonstrate concept. The concept is the use of bacteria as an enzyme pool in a chemical sense rather than as a biological growth system. The difference between the two approaches is principally one of time, since the concept allows the use of micro amounts of reagents in the absence of a requirement for growth and under conditions allowing the achievement of a high level of stability and reproducibility.

Earlier we discussed some of the characteristics of conventional media systems which are recognized as deficient in one or more parameters. What happens when precise amounts of specific substrates and detection reagents are applied to paper in micro amounts, separated from each other when necessary, and the total system for measuring a chemical event is maintained before use at less than 3% moisture?

1. Since growth is not a factor in measuring a biochemical event, data in all reagent-impregnated test strip systems can be obtained in 4 hours or less.

2. Since equipment has been developed that will accurately apply specific volumes of reagents per unit area of paper, paper strip systems are highly reproducible as well as precise.

3. Since all reagents applied to paper are dried to contain less than 3% moisture and incompatible reagents are separated when necessary, test strip systems are stable for at least 2 years at 4°C (1 year—ONPG) when kept dry and protected from light.

4. In our laboratories and in many controlled clinical studies, accuracy of this system has been shown to be of a high order. Publications from independent investigators will appear in the literature in the near future.

5. Reagent-impregnated test strips measure one biochemical parameter per strip. Therefore, the system maintains design flexibility in identification procedures. In many cases all 12 biochemical parameters will not be needed for identification. In some cases more than the 12 parameters presently adapted to strip systems will be required. In all cases full flexibility is maintained.

6. In many laboratories the media storage problem precludes maintainance of an adequate variety of test systems. Application of reagents to paper eliminates significant storage problems.

7. Since all reagents (except KOH) for all test systems are impregnated onto each strip, maintainance and preparation of unstable reagents is eliminated.

8. All reactions are easily interpretable against a white paper background.

Data generated with the use of reagent-impregnated test strips are shown in Tables 1, 2, and 3. All bacterial cultures were fresh clinical isolates belonging to the family *Enterobacteriaceae*. Cultures were identified by one technologist using conventional media representing the same tests as those available on reagent-impregnated test strips plus other tests where required. A second technologist used the test strips; data were recorded separately and the data compared. In each case, a given reaction on conventional media was judged to be definitive.

Table 1 shows test result agreement between PathoTec and conventional biochemical procedures. Four tests (cytochrome oxidase, phenylalanine deaminase, ornithine decarboxylase, and ONPG) yielded complete correlation; seven tests (malonate, nitrate reductase, indole, Voges-Proskauer, lysine decarboxylase, esculin hydrolysis, and H_2S) yielded a correlation greater than 97%. The urease test strip gave a correlation of 92.5%. Table 2 shows the efficacy of the system in identifying *Enterobacteriaceae*. Identification accuracy of ten biochemical tests was 93.7%; with the addition of ornithine decarboxylase and ONPG, this level increased to 97%. Table 3 is a list of organisms which were misidentified or not identified and the reason for error.

These are the concepts involved in paper strip technology. To date, 12 conventional biochemical tests have been adapted to strip technology, and others are under study. We believe that this new approach to microbiology will be of significant utility to the laboratory.

Since the presentation of this paper in June 1973, several reports have appeared in the literature:

1. R. Rosner, Evaluation of the PathoTec Rapid I-D System and Two Additional Experimental Reagent-Impregnated Test Strips, *Appl. Microbiol.* **26** (6), 890–893 (1973).

2. D. J. Blazevic, P. C. Schreckenberger, and J. M. Matsen, Evaluation of the PathoTec Rapid I-D System, *Appl. Microbiol.* **26** (6), 886–889 (1973).

3. S. C. Edberg, V. Gold, M. Novack, H. Slater, and J. M. Singer, A Direct Inoculation Procedure for the Rapid Identification of Enterobacteriaceae from Blood Culture, *Proc. A. S. M.*, paper #M313 (1974).

4. P. C. Schreckenberger and D. J. Blazevic, The Use of Rapid Biochemical Tests in the Identification of Anaerobic Bacteria, *Proc. A. S. M.*, paper #M339 (1974).

Table 1 Test Result Agreement between PathoTec and Conventional Biochemical Procedures

Test	Number Agreeing/Number Tested	Percentage Agreement	False Negative Strip Reactions	Organisms	False Positive Strip Reactions	Organisms
Cytochrome oxidase	333/333	100	None		None	
Phenylalanine deaminase	333/333	100	None		None	
Ornithine decarboxylase	333/333	100	None		None	
ONPG	333/333	100	None		None	
Malonate utilization	331/333	99.4	2	*Enterobacter*	None	
Nitrate reduction	331/333	99.4	2	*P. morganii*	None	

Test	Fraction	%	No.	Organism	No.	Organism
Indole production	330/333	99.1	2	*P. vulgaris*	1	*Citrobacter*
Voges-Proskauer	330/333	99.1	2	*Klebsiella*	1	*Klebsiella*
Lysine decarboxylase	330/333	99.1	2	*S. marcescens* and *S. liquefaciens*	None	
			1	*Klebsiella*		
Esculin hydrolysis	325/333	97.6	1	*Enterobacter*	1	*Escherichia*
			3	*Serratia*	1	*Citrobacter*
			1	*Klebsiella*	1	*P. morganii*
H₂S detection	324/333	97.2	2	*P. mirabilis*	6	*P. morganii*
			1	*Salmonella*		
Urease production	308/333	92.5	8	*Citrobacter*	5	*Klebsiella*
			7	*Klebsiella*	2	*E. aerogenes*
			3	*E. cloacae*		

3275/3330 = 98.3% (10 tests)

3941/3996 = 98.6% (12 tests)

Table 2 Strains tested and Agreement of Identification between Conventional Media and PathoTec "Rapid I-D System" with and without the Additional Strips

Organism	Number of Strains Tested	PathoTec Rapid I-D Number Correct/Number Tested	Percentage	Rapid I-D + OD & ONPG Number Correct/Number Tested	Percentage
Shigella	20	20/20	100	20/20	100
Edwardsiella	18	18/18	100	18/18	100
Arizona	24	24/24	100	24/24	100
Providencia	21	21/21	100	21/21	100
P. rettgeri	9	9/9	100	9/9	100
P. mirabilis	25	25/25	100	25/25	100
Escherichia	23	22/23	95.6	22/23	95.6
Salmonella	21	20/21	95.2	20/21	95.2
E. aerogenes	27	26/27	96.3	27/27	100
E. cloacae	20	19/20	95.0	20/20	100
Citrobacter	16	15/16	93.8	15/16	93.8
Klebsiella	20	18/20	90.0	19/20	95.0
P. vulgaris	20	18/20	90.0	20/20	100
S. marcescens and *S. liquefaciens*	48	45/48	93.8	45/48	93.8
P. morganii	21	12/21	57.1	17/20	85.0
TOTAL	333	312/333	=93.7	323/333	97.0

Table 3 Organisms Misidentified or Not Identified by PathoTec "Rapid I-D System" and Effect of Adding OD and ONPG Strips

Conventional Identification	PathoTec System Identification	Number of Isolates Misidentified		Disagreement Resulting in Misidentification	Number of Isolates Misidentified After Addition of OD and ONPG
Escherichia	Not identified	1		Esculin strip positive	1
Salmonella	Not identified	1		H_2S strip negative	1
Citrobacter	Not identified	1		Indole & esculin strips positive	1
Klebsiella[a]	*Escherichia*	1		Esculin strip negative	1
Klebsiella[b]	*Enterobacter*	1		Urease strip negative	0
E. aerogenes[c]	*Klebsiella*	2		Urease strip positive	0
S. marcescens[d]	Not identified	3		Esculin strip negative	3
P. vulgaris	*P. mirabilis*	2		Indole strip negative	0
P. morganii	Not identified	2		Nitrate strip negative	2
P. morganii	*Proteus sp.*	1		Esculin strip positive	1
P. morganii	*P. vulgaris*	6		H_2S strip positive	0
TOTAL		21	(6.3%)		10 (3.0%)

[a]VP and malonate negative strain.
[b]Motile strain.
[c]Nonmotile strains.
[d]VP negative strains.

CHAPTER 23

Enterotube Roche—A Rapid and Accurate Method for the Identification of Enterobacteriaceae

RUDOLF GALLIEN

SUMMARY

A new, improved Enterotube has been developed to eliminate the major problem with lysine decarboxylase in the previous product. Now a new arrangement of the improved media in the different chambers, especially the additional uptake of an ornithine decarboxylase compartment, permits the rapid identification of 5 groups and 17 members of the family Enterobacteriaceae in one tube.

Compared with conventional methods, the advantages of this useful device for the bacteriological routine laboratory are: (*a*) complete arrangement of routinely used media, (*b*) simplicity in use and reading, (*c*) shortened procedure time, and (*d*) marked savings in time devoted to media preparation, glass washing, and sterilization, and in raw materials, media, and storage space. The revised Enterotube provides excellent agreement with conventional methods and media.

INTRODUCTION

One year ago the first type of Enterotube Roche appeated on the European market. This tube, was a "ready-to-use" test system for the routine identification of Enterobacteriaceae. It consisted of eight separate compartments, containing eight prepared selective media, which were inoculated simultanously by drawing out the central needle. This elegant, quick, and safe inoculation method avoids contamination of the media, as well as errors inherent in multiple-tube techniques. After incubation overnight, nine biochemical properties of the inoculated rod served for its identification:

Medium	Reaction Detected
Citrate	Citrate utilization
Lysine	Lysine decarboxylase
Lactose	Acid production
Dulcitol	Acid production
Urea	Urea cleavage
Phenylalanine	Phenylalanine deaminase
H_2S—indole	H_2S and indole formation
Dextrose	Acid production

To detect the indole reaction, 3 drops of Kovacs reagent are injected into the H_2S-indole compartment. A redding of the nearly colorless reagent indicates a positive reaction. Also, the phenylalanine deaminase reaction could be read only after addition of 3 to 4 drops of 10% ferric chloride solution through the cellophane covering.

All experience suggests that this type of Enterotube is a simple, rapid, and useful device in the bacteriological routine laboratory. The major advantage of the test system is the possibility of simultaneous inoculation from a single colony.

Discrepancies between Enterotube and the conventional methods are noted in the results of the lysine decarboxylase test. Thus agreement in the lysine decarboxylase reaction of Enterotube with conventional tests in only 87.6% of strains could be observed, whereas all the other tests showed a reliability between 95.0 and 98.8% (1). These false-positive reactions of lysine decarboxylase, expecially found with *Shigella* and *Proteus* strains (2), and unsufficient differentiation of the members of the *Klebsiella–Enterobacter–Serratia* group were reason enough to develop an improved Enterotube giving excellent agreement in all reactions with conventional methods.

389

IMPROVED ENTEROTUBE

The improved type of Enterotube, now available in Europe is no different with respect to the empty compartment system from the former type, although there is one more aeration hole in the side of the tube. But the arrangement of the media has been changed markedly. Now 8 media permit the detection of 11 biochemical reactions of Enterobacteriaceae, two more than in the first type.

Medium	Reaction detected
Citrate	Citrate utilization
Urea	Urea cleavage
PA/Dulcitol	Phenylalanine deaminase or acid production
Lactose	Acid production
H$_2$S-Indole	H$_2$S and indole formation
Ornithine	Ornithine decarboxylase
Lysine	Lysine decarboxylase
Dextros–gas	Acid and gas production

New in the media combination is the ornithine decarboxylase compartment. The uptake of this important reaction became possible by combining the phenylalanine deaminase reaction (PA) with the dulcitol reaction within one compartment, since all *Proteus* species, which always are PA positive, are coincidentally dulcitol negative.

The next improvement appears in the incorporation of ferric ammonium citrate into the PA/Dulcitol medium, which eliminates the need to add the 10% ferric chloride solution after incubation. Also advantageous is the possibility of detecting the gas reaction in the dextrose chamber. The wax overlay, which can be found in the lysine and ornithine chambers too, ensures anaerobic conditions for dextrose fermentation, which leads to gas formation and lifting of the wax overlay from the surface of the agar. The wax overlay in the lysine and ornithine compartments prevents false-positive reactions. An additional amendment of the lysine decarboxylase reaction is attributed to the elimination of lactose, which prevents confusion by lactose fermenters.

The handling of the improved Enterotube is unchanged with one exception. After withdrawing the needle through all eight compartments, the user must reinsert the needle through the dextrose, lysine, and ornithine compartments, such that the tip of the needle can be seen in H$_2$S-Indole compartment. At a notch, the needle is broken by bending. The remaining part of the needle, together with the wax overlay, ensures anaerobic conditions.

EVALUATION OF THE IMPROVED ENTEROTUBE ROCHE

With the new Enterotube system the user can expect a high degree of accuracy in the identification of Enterobacteriaceae. If Enterotube is used as recommended by the manufacturer, it represents a proper, acceptable alternative to conventional methods of biochemical identification. Of course, Enterotube cannot replace serologic tests for final identification, especially of *Salmonellae*.

Three papers recently published in the United States confirm the high value of the revised Enterotube for the bacteriological laboratories. The main value can be seen in the overall agreement with conventional methods with 95.2% (3), 97.2% (4), and 64.4% (5) of the cultures. The percentage of correctly identified cultures is mentioned as 96.4% (6) and 98.6%. In addition, the urea and citrate media in the Enterotube seem to be more sensitive than conventional media.

All authors emphasize that the advantages offered by Enterotube include simplicity in use; convenient, quick inoculation of 8 compartments with 11 biochemical reactions; assurance of always having stable, sterile, and controlled devices; and economy possible in sparing laboratory factilities. Three different methods for reading and evaluation of the incubated tubes are very helpful for the user and ensure the results additionally.

REFERENCES

1. Harry R. Elston, Judith A. Baudo, Joann P. Stanek, and Mercedes Schaab, Multi-Biochemical Test System for Distinguishing Enteric and Other Gram-Negative Bacilli, *Amer. Soc. Microbiol.*, **22**, 408–414 (1971).

2. H. Knothe and W. -D. Strohm, Biochemische Differenzierung von Enterobacteriaceae mit einem Multiagarsystem, *Diagnostik*, **6**, 31–34 (1973).

3. K. M. Tomfohrde et al., Evaluation of the Revised Enterotube, presented at a meeting of the American Society for Microbiology, Philadelphia, April 23–28, 1972.

4. H. E. Morton, and M. A. J. Monaco, Comparison of Enterotubes and Routine Media for the Identification of Enteric Bacteria, *Amer. J. Clin. Pathol.*, **56**, 64 (1971).

5. W. -D. Leers, and Kay Arthurs, Routine Identification of Enterobacteriaceae with the "Improved Enterotube." Department of Medical Microbiology, University of Toronto and The Wellesley Hospital, Toronto.

6. K. M. Tomfohrde, D. L. Rhoden, P. B. Smith, and A. Balows, Evaluation of the Redesigned Enterotube—A System for the Identification of Enterobacteriaceae, *Appl. Microbiol.*, **25**, 301–304 (1973).

CHAPTER 24

Evaluation of Different Diagnostic Kits for Enterobacteriaceae

CARL–ERIK NORD, TORKEL WADSTRÖM, ANN DAHLBÄCK

INTRODUCTION

Identification of gram-negative bacteria of the family Enterobacteriaceae from clinical specimens has always been an important practical problem for the clinical laboratory. Attempts to reach not just a group but a species diagnosis are justified by the increased prevalence of Enterobacteriaceae in specimens from hospitals. Accurate diagnosis is helpful for the surveillance of nosocomial infections and spread of antibiotic resistance by R factors (1–5). However, it is evident that the handling of many clinical cultures is not only time-consuming but also is costly in terms of media preparation and equipment.

Methods that would reduce the routine work and equipment necessary for accurate identification of Enterobacteriaceae would be valuable in the clinical laboratory. Recently different multitest systems such as PathoTec (7–13), Enterotube (11, 14–17), R/B (18–23), API (24–29), and AuxoTab (30,31) have been described. For the most part, close correlations were reported between the results obtained using these multitest systems and those obtained by standard biochemical methods. In this study the five diagnostic kits were tested and compared with one another and with standard biochemical tests in relation to the accuracy and practicability of each test kit.

MATERIALS AND METHODS

Strains Tested

Twenty stock cultures (Table 1) and 161 fresh clinical isolates from specimens (Table 2) submitted to the Swedish National Bacteriological Laboratory for identification were used to evaluate the different test kits. Each strain was assigned a code number, and the identity of the strains verified by standard biochemical tests (see below) was kept unknown until the study was completed. Each culture was streaked onto a CLED agar plate (Oxoid) and an Endo agar plate (Difco). The media employed for the standard tests were those recommended by Cowan and Steel (32), Martin (33), and Edwards and Ewing (34). Serology was not used in this study, but many cultures were also serologically identified at the National Bacteriological Laboratory. All nomenclature and taxonomic identifications were made according to Edwards and Ewing (34).

Table 1 Microorganisms Used in the Evaluation of the Five Test Systems

Microorganism	Number
Escherichia coli	22
Klebsiella pneumoniae	18
Klebsiella pneumoniae var. oxytoca	3
Klebsiella ozeanae	2
Enterobacter aerogenes	2
Enterobacter cloacae	8
Enterobacter hafniae	2
Enterobacter liquefaciens	9
Serratia marcescens	3
Citrobacter freundii	18
Proteus mirabilis	21
Proteus vulgaris	20
Proteus rettgeri	5
Proteus morganii	17
Providencia stuartii	6
Providencia alcalifaciens	1
Salmonella	10
Shigella	10
Edwardsiella tarda	1
Arizona	5

Table 2 Sources of Microorganisms

Source	Number
Clinical urine	115
Sputum	15
Stool	22
Wound	7
Ear	2
Stock culture	20

Description of the Diagnostic Kits

Enterotube® (Improved)

The molded plastic tube called Enterotube by the manufacturer is divided into eight compartments, each one containing a slant of a standard biochemical medium. A wire that serves as an inoculator extends lengthwise through the tube and projects from either end. The following 11 biochemical tests are used: acid and gas from glucose, hydrogen sulfide, indole, phenylalanine deaminase, urease, dulcitol, lactose, ornithine decarboxylase, lysine decarboxylase, and citrate. Incubation was at 37°C for 24 hours. The biochemical reactions were read according to the manufacturer's directions (Hoffman–La Roche).

PathoTec®

The PathoTec reagent systems are dry paper strips impregnated with reagents that detect the presence of specific enzymes or metabolic end products. The following tests are available: oxidase, nitrate reduction, phenylalanine deaminase, urease, indole, hydrogen sulfide, lysine decarboxylase, acetoin, malonate utilization, and esculin hydrolysis. All tests except oxidase were incubated at 37°C for 4 hours. Tests were read according to the manufacturer's instructions (General Diagnostics).

API

Each test in Analytab Products's minimized system, API, is performed within a sterile plastic tube; the tube contains the appropriate substrates and is affixed to an impermeable plastic packing. The 20 tests consist of the following: ONPG, arginine dihydrolase, lysine decarboxylase, ornithine decarboxylase, citrate, hydrogen sulfide, urease, tryptophan deaminase, indole, acetoin, gelatin and fermentation tests of glucose, mannitol, inositol, sorbitol, rhamnose, sucrose, melibiose, amygdalin, and arabinose. Incubation was at 37°C for 24 hours. All tests were performed as recommended by the manufacturer.

(Modified) R/B Tubes

These modified R/B tubes are manufactured by Diagnostic Research Inc., as the "two-tube system." The first tube contains appropriate media for recognizing activities such as phenylalanine deaminase, lactose fermentation, gas production from glucose, lysine decarboxylase and hydrogen sulfide production. The second tube is used to study reactions such as motility, indole production, and ornithine decarboxylase. The manufacturer's instructions for inoculation, loose capping, and incubation (37°C for 24 hours) were followed.

Auxotab®

The Auxotab Enteric I cards were supplied by Colab Laboratories, Inc., and were used as the manufacturer recommended. A colony from an agar plate was transferred to 5 of ml brain-heart infusion and incubated at 37°C for 3.5 hours. The broth culture was then centrifuged at 3500 rpm for 15 minutes, and the cells were suspended in 1.8 ml of distilled water, pH 6.7. The cell suspension was inoculated into the upper opening of each capillary on the card and incubated at 37°C for 1 hour. Next the viability control (resazurin) was examined, and if the reaction was positive the card was incubated for an additional 2 hours at 37°C. The system consists of a card with 10 capillary units containing the following reagents: viability control (resazurin reduction), malonate, phenylalanine deaminase, hydrogen sulfide, sucrose, o-nitrophenyl-β-D-galactopyranoside, lysine decarboxylase, ornithine decarboxylase, urease, and tryptophan (indole).

RESULTS AND DISCUSSION

One hundred eighty-one cultures were tested thus by the Enterotube, PathoTec, API, R/B, and Auxotab systems, and by standard biochemical assays.

Enterotube

Table 3 shows the agreement of results of 11 tests in the Enterotube and their corresponding standard tests. Acid from glucose, dulcitol, H_2S production, ornithine decarboxylase, and urease production had better than 99% agreement in the parallel tests. For lactose, lysine decarboxylase, and indole, the agreement was slightly less than the tests just mentioned. The agreement for citrate utilization, gas from glucose, and production of phenylalanine deaminase was slightly under 90%.

Morton and Monaco (15) compared the Enterotube and routine media for identification of 147 clinical cultures belonging to *Escherichia*, *Enterobacter*, *Citrobacter*, *Klebsiella*, *Proteus*, *Providencia*, *Pseudomonas*, *Salmonella*, and *Shigella*. They reported an agreement in the results obtained by two methods of 82.3%. On the other hand, Elston et al. (16) found no differences in the indole test and only small discrepancies were noted in the citrate test. Lysine decarboxylase was stated to be an unacceptable alternative to the standard methods. Martin et al. (14) also found that the lysine decarboxylase test was unsatisfactory, but reported good agreement with hydrogen sulfide, indole, citrate, glucose, and lactose. Less than 85% agreement was obtained with dulcitol, urease, and phenylalanine deaminase. Tomfohrde et al. (17) also found this disagreement between the Enterotube and the conventional system for citrate. However, the redesigned Enterotube has been reported to have a better agreement with conventional test procedures than the original Enterotube system described by Grunberg et al. (11).

Table 3 Comparative Test Results:
Enterotube and Standard System

Test	Agreement (%)
Dulcitol	100
Ornithine	100
Glucose (acid)	99.4
Hydrogen sulfide	99.4
Urease	99.4
Indole	97.8
Lactose	96.7
Lysine	95.0
Citrate	88.9
Phenylalanine	88.9
Glucose (gas)	88.5

The second part of this analysis concerned the accuracy of the Enterotube in correctly identifying unknown bacteria belonging to Enterobacteriaceae by using the Enterotube numerical coding and identification system for Enterobacteriaceae. There was good agreement between this coding system and the identification scheme of Edwards and Ewing (34) in correctly placing organisms into genera or groups. One strain typed as *P. alcalifaciens* in the conventional scheme was typed as *P. vulgaris* in the Enterotube coding system. One strain of *K. ozeanae* could not be identified with the coding system.

From our tests and the literature it appears that the Enterotube will accurately identify between 80 and 100% of the enteric bacteria, and it seems to offer an acceptable alternative to other systems used for identification of strains belonging to Enterobacteriaceae. It is easy to use, reliable in the identification, readily disposable, and easily stored.

PathoTec

The results of the biochemical tests of the PathoTec system and the standard system are listed in Table 4. Agreement between the two methods was 100% in the following tests: oxidase, malonate, nitrate, and esculin. Urease, indole, acetoin, and hydrogen sulfide showed about 98% correlation; lysine decarboxylase showed 95% and phenylalanine deaminase, 91%. Matsen and Sherris (12) found a good correlation with oxidase, acetoin, and urease, and disagreement with the lysine-decarboxylase test. Grunberg et al. (11) also reported a rather good correlation of the Enterotube and the Pathotec tests in the following tests: indole, phenylalanine deaminase, and urease. However, the best correlation between the paper strips and the standard media was reported by Rosner (13), who found a 99.3% correlation between the results. The PathoTec system offers an identification system giving results in good agreement with the conventional procedures. It also allows the user to choose tests, and identification of the strains can be made within 4 hours from primary isolation.

Table 4 Comparative Test Results: PathoTec and Standard System

Test	Agreement (%)
Oxidase	100
Nitrate	100
Malonate	100
Esculin	100
Urease	98.3
Indole	97.8
Acetoin	97.8
Hydrogen sulfide	97.8
Lysine	95.0
Phenylalanine	91.1

API

Table 5 shows the agreement of the results of the 20 tests in the API system and their corresponding standard tests. The results of the 20 paired tests showed good agreement of the order of 95 to 100%. Citrate was the only test that was under 95% agreement, and this was also recently reported by Smith et al. (29). Both Washington et al. (25) and Smith et al. (29) found that the conventional system detected more urease positive cultures than the API system. This result was not confirmed in our study. Interestingly, Gardner et al. (28) were able to identify only 25% of the enteric bacteria with the API system in a clinical laboratory. They concluded that the system was not a reliable substitute for the standard biochemical identification methods with tube media—again, a result unconfirmed in our study.

Table 5 Comparative Test Results: API and Standard System

Test	Agreement (%)
Glucose	100
Sucrose	100
Rhamnose	100
Arabinose	100
Melebiose	100
Amygdalin	100
Hydrogen sulfide	99.4
Indole	99.4
ONPG	99.4
Tryptophan	99.4
Mannitol	98.9
Lysine	98.9
Arginine	98.9
Inositol	98.3
Sorbitol	98.3
Gelatin	97.8
Urease	97.2
Acetoin	97.2
Ornithine	95.6
Citrate	86.7

To test the identification register called the API profile register, we compared it with the identification scheme of Edwards and Ewing (34). Very good agreement was obtained. Only one bacterium typed as *Enterobacter cloacae* according to Edwards and Ewing, was missed. This specimen was called *Salmonella* in the API profile register. Thus the API system gives an opportunity to apply the Adansonian concept of numerical taxonomy (36) directly in clinical laboratory work. At the present time the profile register contains the data of more than 25,000 strains tested with the API system in various parts of the world. The API system would suit a modern computerized clinical laboratory. In summary, the API system is easy to use, accurate in identifying unknown cultures, and economical in terms of the information provided.

R/B System

Table 6 compares results from the eight tests of the R/B system with the results from the same eight tests performed in our standard media. As indicated in the table, the agreement of tests was good. In a previous study, Smith et al. (20) reported that the motility test and the lysine decarboxylase test were unreliable. These tests have now been modified and were found to be reliable in a new study (21). Martin et al. (19) found that the reactions of the R/B system were more difficult to read and interpret than those of the conventional system; they concluded that the R/B system was not an acceptable alternative to the conventional methods used. Isenberg and Painter (22), on the other hand, considered that the modified R/B system performed well in primary identification of various species belonging to Enterobacteriaceae. The modified system tested by McIlroy et al. (23) gave reactions comparable to the conventional tests. These reactions were hydrogen sulfide, lysine decarboxylase, ornithine decarboxylase, and glucose. The findings obtained in different studies indicate that the R/B system in its recently modified form represents an alternative approach for the identification of Enterobacteriaceae.

Table 6 Comparative Test Results:
R/B System and Standard System

Test	Agreement (%)
Ornithine	100
Hydrogen sulfide	100
Motility	100
Lysine	99.4
Glucose	98.9
Lactose	98.9
Phenylalanine	98.3
Indole	97.2

Auxotab

The evaluation of the accuracy and convenience of the Auxotab Enteric I system for rapid identification of Enterobacteriaceae was first tested by incubation for 3.5 hours in brain-heart infusion according to the manufacturer's instructions. However, many strains failed to achieve the minimal concentration of organisms to complete the process within 3.5 hours. Therefore, in another experiment the strains were incubated for 18 hours and the Auxotab process started next day. The biochemical reactions obtained by standard means are compared with those obtained in the Auxotab system after 3 hours' incubation at 37°C in Table 7. However, poor agreement was noted between the Auxotab hydrogen sulfide and its conventional counterpart. There was between 86 to 88% agreement between the Auxotab ornithine decarboxylase, ONPG, and indole tests with their standard counterparts. Sucrose, urease, lysine decarboxylase, and phenylalanine deaminase showed 92 to 95% agreement, and malonate was the only test with full agreement. The Auxotab system in its present form is tedious and inconvenient. It is also doubtful whether this system is rapid compared with the other systems, since it takes 18 hours to complete the biochemical reactions (30). The obvious need for modification of this product was pointed out recently by Rhoden et al. (31).

Table 7 Comparative Test Results: Auxotab System and Standard System

Test	Agreement (%)
Malonate	100
Phenylalanine	95.0
Lysine	93.9
Urease	92.1
Sucrose	92.1
Indole	88.9
ONPG	88.9
Ornithine	86.7
Hydrogen sulfide	80.0

After this short evaluation of the different kits, it is not possible to recommend the best set, because this depends on many factors. All systems have innate advantages and disadvantages. If there are many tests in the system, more information about the investigated strains is obtained. The identified microorganisms can be placed into genera, groups, or species, depending on the number of tests used. However, it is also important to evaluate the single test in the system used. In Sweden, for example, the primary identification of gram-negative rods belonging to Entero-bacteriaceae has often been carried out in routine clinical bacteriological work with a set of three biochemical tests—the methyl red, acetoin, and citrate utilization tests. However, these tests are time-consuming; for example, the methyl red test should never be read until the cultures have been incubated for at least 2 days. Sometimes the species of an isolate cannot be determined until the third day after the bacteriological specimen has arrived in the laboratory. On the other hand, the result of antibiotic sensitivity testing is usually obtained on the second day. From this point of view it must be more satisfactory to have the bacteria speciated at the same time the result of antibiotic sensitivity testing is obtained. Therefore, we have investigated 14 biochemical tests that could be used for the identification of different species belonging to Enterobacteriaceae within 24 hours. The following tests were investigated: fermentation of adonitol, mannitol, lactose, and sucrose; production of β-galactosidase; determination of arginine di-hydrolase and lysine and ornithine decarboxylases; production of hydrogen sulfide; motility; production of indole; urease activity; citrate utilization; and test for acetoin.

The best results were obtained with the following tests: ONPG, acetoin, indole, urease, hydrogen sulfide, and ornithine decarboxylase. We have now used these six biochemical reactions for one year at our laboratory for the primary identification of Enterobacteriaceae from clinical urine specimens. Our results have been good, but the preparation and control of these standard biochemical media is time-consuming. Therefore, their use may be limited, especially in small diagnostic laboratories. The commercial tests thus offer a good alternative.

It is also important to discuss which classification is to be used. All the kits tested are based on the nomenclature of Edwards and Ewing (34). Nomenclature and microbiological tests recommended by Cowan and Steel (32) and Kauffmann (35) are used at the Swedish National Bacteriological Laboratory for identification of strains belonging to Enterobacteriaceae. It is also possible to use the nomenclature and tests recommended by the eighth edition of *Bergey's Manual*, or other diagnostic schemes. There is also a need for continuous re-evaluation of these diagnostic kits, because of the tendency of the firms to change their tests quite often—redesigning, improving, and reimproving kits. For future surveys of the accuracy of test kits, it is essential to have more detailed information from the manufacturers about the media composition, batch number, date of preparation, and storing time.

The commercial test systems are easy to use, accurate in the identification of unknown cultures, easily stored, and economical. At present, the Enterotube, PathoTec, API, and R/B systems offer an alternative to standard biochemical tests for identifying Enterobacteriaceae in a clinical laboratory, in conjunction with information gained from observing colony morphology and growth on selective media, and from results of serological and other tests.

REFERENCES

1. T. Watanabe, *Bacteriol Rev.*, **27**, 87 (1963).
2. D. H. Smith and S. E. Armour, *Lancet*, **2**, 15 (1966).
3. P. Gardner and D. H. Smith, *Ann. Intern. Med.*, **71**, 1 (1969).
4. S. Mitsuhashi, *Transferable Drug Resistance Factor R*, University Park Press, Baltimore, London, and Tokyo, 1971.
5. M. Jonsson, *Scand. J. Infect. Dis.*, Suppl. 5, 1972.
6. K. G. Nardan, P. A. M. Grunce, and D. D. A. Morsel, *Antonie van Leeuwenhoek*, **33**, 184 (1967).
7. D. K. Weaver, E. K. H. Lee, and M. S. Leaky, *Amer. J. Clin. Pathol.*, **49**, 494 (1968).
8. K. A. Bouchardt, *Amer. J. Clin. Pathol.*, **49**, 748 (1968).
9. N. N. Small, *Amer. J. Med. Technol.*, **34**, 65 (1968).
10. D. Amsterdam and M. W. Wolfe, *Appl. Microbiol.*, **16**, 1460 (1968).
11. E. Grunberg, E. Titsworth, G. Beskid, R. Cleeland, and W. F. Delorenzo, *Appl. Microbiol.*, **18**, 207 (1969).
12. J. M. Matsen and J. C. Sherris, *Appl. Microbiol.*, **18**, 452 (1969).
13. R. F. Rosner, *Amer. J. Clin. Pathol.*, **54**, 587 (1970).
14. W. J. Martin, P. K. W. Yu, and J. A. Washington, II, *Appl. Microbiol.*, **22**, 96 (1971).
15. H. E. Morton and M. A. Monaco, *Amer. J. Clin. Pathol.*, **56**, 64 (1971).
16. H. R. Elston, J. A. Bando, J. P. Stanek, and M. Schaab, *Appl. Microbiol.*, **22**, 408 (1971).
17. K. M. Tomfohrde, D. L. Rhoden, P. B. Smith, and A. Balows, *Appl. Microbiol.*, **25**, 301 (1973).
18. E. D. O'Donnell, F. J. Kaufmann, E. D. Longo, and P. D. Ellner, *Amer. J. Clin. Pathol.*, **53**, 145 (1970).
19. W. J. Martin, R. J. Birk, P. K. W. Yu, and J. A. Washington, II, *Appl. Microbiol*, **20**, 880 (1970).
20. P. B. Smith, D. L. Rhoden, K. M. Tomfohrde, C. R. Dunn, A. Balows, and G. J. Hermann, *Appl. Microbiol.*, **21**, 1036 (1971).
21. P. B. Smith, K. M. Tomfohrde, D. L. Rhoden, and A. Balows, *Appl. Microbiol.*, **22**, 928 (1971).
22. H. D. Isenberg and B. G. Painter, *Appl. Microbiol.*, **22**, 1126 (1971).
23. G. T. McIlroy, P. K. W. Yu, W. J. Martin, and J. A. Washington, II, *Appl. Microbiol.*, **24**, 358 (1972).

24. J. Buissière and P. Nardon, *Ann. Pasteur Inst.*, **115**, 218 (1968).

25. J. A. Washington, II, P. K. W. Yu, and W. J. Martin, *Appl. Microbiol.*, **22**, 267 (1971).

26. F. N. Guillerment and A. M. B. Desbresles, *Rev. Inst. Pasteur, Lyon*, **4**, 71 (1971).

27. B. B. Nielsen, *Med. Dansk Dyraelegefören.*, **54**, 951 (1971).

28. J. M. Gardner, B. A. Snyder, and D. Gröschel, *Pathol. Microbiol.*, **38**, 103 (1972).

29. P. B. Smith, K. M. Tomfohrde, D. L. Rhoden, and A. Balows, *Appl. Microbiol.*, 24, 449 (1972).

30. J. A. Washington, II, P. K. W. Yu, and W. J. Martin, *Appl. Microbiol.*, **23**, 298 (1972).

31. D. L. Rhoden, K. M. Tomfohrde, P. B. Smith, and A. Balows, *Appl. Microbiol.*, **25**, 284 (1973).

32. S. T. Cowan and K. J. Steel, *Manual for the Identification of Medical Bacteria*, Cambridge University Press, London, 1970.

33. W. J. Martin, "Enterobacteriaceae," in *Manual of Clinical Microbiology*, J. E. Blair, E. H. Lenette, and J. P. Truant., Eds., American Society for Microbiology, Bethesda, Md., 1970, pp. 151–174.

34. P. R. Edwards and W. H. Ewing, *Identification of Enterobacteriaceae*, Burgers, Minneapolis, Minn., 1972.

35. F. Kauffmann, *The Bacteriology of Enterobacteriaceae*, Munksgaard, Copenhagen, Denmark, 1969.

36. W. R. Lockhart and J. Liston, *Methods for Numerical Taxonomy*, American Society for Microbiology, Bethesda, Md., 1970.

CHAPTER 25

Evaluation of the API, the PathoTec, and the Improved Enterotube Systems for the Identification of Enterobacteriaceae

R. S. MOUSSA

ABSTRACT

The API system, the PathoTec system, and the improved Enterotube system have been compared with the conventional methods in the examination of 140 cultures belonging to the genera *Escherichia, Shigella, Salmonella, Arizona, Citrobacter, Klebsiella, Enterobacter,* and *Proteus.* The improved Enterotube system gave the best correlation with the conventional methods. The API and the PathoTec systems were less satisfactory. Moreover, the improved Enterotube system was simpler, required less work, and was easier to handle than the other two methods. Its tests have been well chosen to permit correct identification of most Enterobacteriaceae organisms.

INTRODUCTION

Washington, Yu, and Martin (10) estimated that 90% of the bacterial cultures that must be identified in clinical laboratories are Enterobacteriaceae. The family Enterobacteriaceae comprises important pathogens belonging to such genera as *Salmonella*, *Shigella*, *Escherichia*, and *Klebsiella*; it is, therefore, important to have at our disposal simple and rapid methods for the identification of isolates belonging to these genera. During recent years, methods of two types have been advocated. On the one hand, there are those which are characterized by the use of a heavy suspension of the pure culture to be identified. The suspension is brought into contact with dehydrated substrates, either in miniaturized cupules (e.g., as in the API system) or absorbed on filter paper (e.g., as in the PathoTec system). On the other hand, there are the methods in which the pure culture is inoculated into a set of small tubes or compartments that contain ready prepared test media (e.g., as in the Enterotube system).

Both types of system are attractive in that they obviate the need to prepare and to store a large number of special media. Certain of these systems are easy to handle, often involving little work or time. For a discussion of the development of this kind of method until 1968, see Hartman (4). In this chapter, we report on tests of the performance of the API, the PathoTec, and the improved Enterotube systems in the characterization and identification of 140 cultures belonging to the Enterobacteriaceae family.

MATERIALS AND METHODS

Cultures

We used 140 Enterobacteriaceae cultures belonging to the genera *Escherichia* (10 strains), *Shigella* (6), *Salmonella* (94), *Arizona* (3), *Citrobacter* (3), *Klebsiella* (11), *Enterobacter* (3), and *Proteus* (10).

The *Shigella* cultures were obtained from the National Collection of Type Cultures (NCTC), London (*Shigella flexneri*, NCTC 3; *Shigella dysenteriae*, NCTC 2966; and *Shigella sonneii*, NCTC 9212) and the Swedish State Bacteriological Laboratory (SSBL), Stockholm (*Sh. flexneri*, SSBL type 1; *Sh. sonneii*, SSBL type 1; and *Shigella boydii*, SSBL type 1). All the *Salmonella*

cultures were isolated in our own laboratory and were serologically identified in the Swiss Salmonella Center, Bern. They belonged to 44 serotypes.

The *Arizona* strains were the NCTC 8297 and 6484 and the SSBL
16 : 23 : 34. The *Citrobacter* strains were the NCTC 9750 and 6071 and the
SSBL type 1. The *Klebsiella* strains included the SSBL *Klebsiella penumoniae*
American Type Culture Collection 1 3882. The *Enterobacter* strains were all
Enterobacter aerogenes and were obtained from NCTC (9735 and 10006) and
SSBL. The 10 *Proteus* strains comprised 7 *Proteus mirabilis*, 2 *Proteus rettgeri*,
and 1 *Proteus morganii*. Table 1 summarizes the biochemical reactions of our
140 cultures.

The API System

Pure cultures were grown on a Difco slant of brain-heart infusion agar for
24 ± 2 hours at 37°C. A heavy suspension of each culture was then prepared
in distilled water, and used, as recommended by the manufacturer (BBL), to
inoculate each of the 20 different cupules supplied on the API plastic strip.
The strip was incubated for 24 ± 2 hours at 37°C.

Conventional tests for β-galactosidase, tryptophane deaminase, and melibiose and amygdaline fermentation were not made, since these tests were
assumed to be of secondary importance in the identification of most Enterobacteriaceae. Evaluation of these four tests has been dropped from our
report. A "coder" and a "profile register" can now be obtained to help
identify Enterobacteriaceae with the API system.

The PathoTec Reagent System

The PathoTec system was supplied by Cosmopharm S.A., Zürich. A heavy
loopful of the brain-heart infusion agar culture—or, as in the API system, a
suspension of it in test tubes—was brought into contact with the paper-
impregnated substrates. Examination for lysine decarboxylase required 6
hours of incubation in a water bath at 37°C; phenylalanine deaminase,
urease, indole, Voges-proskauer, and citrate tests required less time either at
37°C or at room temperature. The PathoTec-LD test permits examination
for two different biochemical reactions at the same time—namely, lysine
decarboxylase and β-galactosidase. In this test, a positive reaction is
obtained with an organism having the enzyme lysine decarboxylase only if
lactose is not fermented. The PathoTec-LD test is offered by the manufacturer as a presumptive test for *Salmonella*; we have used it with our lactose
nonfermenting *Salmonella* and *Arizona* cultures only. During recent months,

Table 1 Biochemical Reactions of the 140 Enterobacteriaceae Cultures Examined

Genus (Number of Cultures)

Test or Substrate	Escherichia (10)	Shigella (6)	Salmonella (94)	Arizona (3)	Citrobacter (3)	Klebsiella (11)	Enterobacter (3)	Proteus (10)
Indole	+	–	–	–	2–/1+	–	–	7–/3+
Voges-Proskauer	–	–	–	–	–	+	+	–
Simmons citrate	–	–	+	+	2+/1–	+	+	9+/1–
H₂S (Kligler)	–	–	93+/1–	+	2+/1–	–	–	7+/3–
Urease	–	–	–	–	–	+	2–/1+	+
Gelatin (37°C)	–	–	–	+	–	–	–	7+/3–
Lysine decarboxylase	–	–	+	+	–	+	+	–
Arginine dehydrolase	8+/2–	4–/2+	62+/32–	+	2–/1+	–	–	–
Ornithine decarboxylase	8+/2–	4–/2+	+	+	2–/1+	–	+	8+/2–
Phenylalanine deaminase	–	–	–	–	–	–	–	+
Glucose	+	+	+	+	+	+	+	+
Glucose, gas	+	–	+	+	+	+	+	+
Lactose	+	–	–	–	+	+	+	–
Sucrose	5+/5–	–	–	–	+	+	+	9+/1–
Mannitol	+	5+/1–	+	+	+	+	+	8–/2+
Dulcitol	8+/2–	–	+	–	–	–	–	–
Inositol	–	–	51+/43–	–	–	+	+	8–/2+
Sorbitol	+	5–/1+	+	+	+	+	+	8–/2+
Arabinose	+	5+/1–	92+/2–	+	+	+	+	–
Rhamnose	8+/2–	–	+	+	+	+	+	–

however, an improved PathoTec system was introduced. We could not examine it as it was unavailable and was indeed unknown to us until we finished this work. The new system drops the citrate test and introduces the H_2S, malonate, nitrate, and esculin tests. The lysine decarboxylase test was also redevised. Moreover, the method of examination has been standardized so that all the tests, except the cytochrome oxidase, now require an incubation of 4 hours at 35 to 37°C. An overall agreement of 95% with the conventional methods is claimed.

The Improved Enterotube System

The improved Enterotubes were obtained from Roche Diagnostics, Nutley, New Jersey. The system comprises 11 tests, instead of nine as in the previous one. The improved Enterotube system include tests for ornithine decarboxylase as well as for production of gas from glucose. Moreover, the iron salt has been added to the phenylalinine compartment (thus eliminating the need for adding it after incubation), and the test has been combined with that for dulcitol fermentation in a single compartment. The test for lactose fermentation has been separated from that of lysine decarboxylase, and the tests for H_2S–indole and for urease were improved.

The two end caps of the improved Enterotube were removed, and the center of the colony to be examined was touched with the sterile tip of the built-in needle. The needle was twisted before it was withdrawn, as recommended by the manufacturer, through the improved Enterotube compartments. The needle was broken at notch, the two end caps replaced, the blue side tape stripped off, and the clear band pushed forward to cover the glucose compartment. The improved Enterotubes were incubated on their flat surfaces for 24 ± 2 hours at 37°C. Before recording the results, a few drops of Kovacs reagent were added through the plastic filter of the H_2S–indole compartment.

The Conventional Methods

Examination for sugar fermentation was made using Oxoid's peptone water and Merck's glucose, lactose, mannitol, sucrose, dulcitol, sorbitol, inositol, and rhamnose, and Fluka's arabinose. Examinations for indole, Voges-Proskauer, and urease were made using Oxoid's tryptone water, MRVP medium and urea agar base, and 40% urea solution, respectively. Difco's Simmons citrate agar, nutrient gelatin, and Kliger's iron agar were used to examine for citrate, gelatin liquefication, and H$_2$S formation. Examination for phenylalanine deaminase, arginine dehydrolase, ornithine decarboxylase, and lysine decarboxylase were made using phenylalanine agar and Moeller decarboxylase broth base to which Fluka's arginine, ornithine, and lysine had been added. After inoculation, all media were incubated for 24 ± 2 hours at 37°C.

Recording of Results

All results were recorded after 24 ± 2 hours. It is possible, however, to extend the incubation time of the API and the improved Enterotube systems, along with the conventional methods, for 48 hours or more.

Results were expressed as follows: $+ + +$ for a clearly positive, $+ +$ for a weak positive,? for a doubtful, and $-$ for a negative result. For evaluating the overall accuracy of each test or method, weak positive results were counted as positive, but doubtful results were considered to be negative.

RESULTS AND DISCUSSION

Accuracy of Tests

The results obtained with each of the three systems examined are compared in Table 2 with those obtained with the conventional methods. For the API system, the indole, urease, gelatin, Voges-Proskauer and sucrose, sorbitol, arabinose, rhamnose, glucose, and mannitol fermentation tests showed an accuracy of between 95 and 100%. The arginine dehydrolase and inositol fermentation had an accuracy of between 90 and 95%. The accuracy of the H$_2$S, lysine decarboxylase, ornithine decarboxylase, and citrate tests was less satisfactory. Discrepancies between the latter tests and the conventional methods were also reported by Washington et al. (10) and Smith et al. (8).

The overall accuracy of the API system (2240 tests) was 92.2%. However, if clearly positive but not weak positive reactions are considered, the overall accuracy of the system drops to 82.7%. Both figures are inferior to the 96.5% reported by Smith and his coworkers (8).

For the PathoTec system, the indole, Voges-Proskauer, phenylalanine deaminase, and urease tests gave an accuracy of 90 to 100%; the citrate and lysine decarboxylase tests were unsatisfactory. Discrepancies between the lysine, urease, and citrate tests have been reported by Matsen and Sherris (7), Grunberg et al. (3), and Martin et al. (6). To improve the performance

Table 2 Agreement Between the API, the PathoTec, and the Improved Enterotube Systems Results and the Conventional Methods

Test or Substrate	API Agreement[a] with Conventional Method	%[b]	PathoTec Agreement with Conventional Method	%	Improved Enterotube Agreement with Conventional Method	%
Indole	138+0	98.6	133+4	97.8	140+0	100
Voges-Proskauer	127+9	97.1	133+4	97.8	110+30	100
Citrate	37+27	45.7	32+37	49.3	98+35	95.0
Hydrogen sulfide	90+34	88.6			138+1	99.3
Urease	136+2	98.6	120+7	90.7	121+19	100
Gelatin	137+0	97.8				
Lysine decarboxylase	78+38	82.8	16+21	38.1[c]		
Arginine dehydrolase	91+41	94.3				

416

Ornithine decarboxylase	95+21	82.8				
Phenylalanine deaminase			130+5	96.4	119+19	98.6
Glucose	120+19	99.3			140+0	100
Glucose, gas					125+15	100
Lactose					108+31	99.3
Sucrose	135+5	100			135+5	100
Mannitol	128+7	96.4				
Dulcitol					121+18	99.3
Inositol	131+0	93.6				
Sorbitol	135+5	100				
Arabinose	139+1	100				
Rhamnose	137+3	100				

[a]Clearly positive + weak positive results.
[b]All positive results together.
[c]For *Salmonella* and *Arizona* cultures only.

of the lysine decarboxylase test, Martin and his coworkers recommended the separation of the lactose and the lysine tests.

The overall accuracy of the PathoTec system (797 tests) was 80.5%. If clearly positive results only are considered, the overall accuracy of the system is only 70.8%.

All the improved Enterotube tests gave an accuracy of 95 to 100%. The least satisfactory was understandably the H_2S test (95.0%); conventional methods are known to vary in detecting its production by most Enterobacteriaceae. Discrepancies between the lysine, urease, indole, citrate, and lactose fermentation tests and the conventional methods have been reported with the previous Enterotube system by Grunberg et al. (3), Elston et al. (1), and Martin et al. (5). Tomfohrde et al. (4) also reported a number of discrepancies between the urease and citrate tests in the improved Enterotube system and the conventional methods. However, 9 of the 15 errors detected by these workers were associated with atypical cultures, and only 6 attributed to the Enterotube. Discrepancies between the ornithine, urease, and dulcitol fermentation tests and the conventional methods were noted in our work. We did not encounter any difficulties in detecting gas production from glucose, except with one *Citrobacter* strain.

The overall accuracy of the improved Enterotube system (1540 tests) was 99.2%. If clearly positive results only are considered, the overall accuracy of the system becomes 88.0%. This corresponds to an overall accuracy of 96.4% reported recently by Tomfohrde and his coworkers (9) and to 94.1% claimed by the manufacturer.

Identification of Enterobacteriaceae

Ewing's classification of the Enterobacteriaceae family (see, e.g., Ref. 2) is based on a combination of 26 essential biochemical reactions. Additional biochemical and serological tests may also be needed for the identification of many genera.

Shortened biochemical schemes necessarily depend on single characteristics in differentiating between most Enterobacteriaceae genera. Although they can be helpful in the tentative identification of many organisms, they are of little use with certain genera, particularly with atypical strains.

An estimation of the number of cultures a system may identify depends not only on the accuracy of the tests it provides but also on the identity of the cultures chosen and the regularity of their characteristics. Most workers, for example, have encountered difficulties in identifying *Enterobacter* cultures by shortened biochemical schemes in general.

The improved Enterotube system, however, seems to be more suited for the identification of most Enterobacteriaceae organisms than the API and the PathoTec systems (Table 3). The improved Enterotube system is simple and requires little work. The time spent in using it and finding out the results is short, compared with the API and the PathoTec systems. Its overall accuracy is high, and its choice of tests has been well selected to permit correct identification of most Enterobacteriaceae organisms.

Table 3 Evaluation of the API, PathoTec, and Improved Enterotube Systems

	System		
Criterion	API	PathoTec	Improved Enterotube
Simplicity of system	Not simple (bacterial suspension prepared)	Not simple (heavy loopful or suspension used)	Simple
Validity of kit if stored as recommended at ±4°C	Long	Indefinite	Short
Work/time spent on test	Much (10 minutes)	Much (15 minutes)	Short (2 minutes)
Time for final result	48 hours	30 hours	24 hours
Overall accuracy of method	92.2%	80.5%	99.2%
Choice of tests for identification of Enterobacteriaceae	Sufficient	Insufficient	Good

REFERENCES

1. H. R. Elston, J. A. Baudo, J. B. Stanek, and M. Schaab, *Appl. Microbiol.*, **22**, 408 (1971).
2. W. H. Ewing, U. S. Public Health Service Publ. 734, Communicable Disease Center, Atlanta, Ga., 1969.
3. E. Grunberg, E. Titsworth, G. Beskid, R. Cleeland, Jr., and W. F. Delorenzo, *Appl. Microbiol.*, **18**, 207 (1969).

4. P. A. Hartman, *Advances in Applied Microbiology*, Suppl. 1, Academic Press, New York, 1968.

5. W. J. Martin, P. K. W. Yu, and J. A. Washington, II, *Appl. Microbiol.*, **22**, 96 (1971).

6. W. J. Martin, S. F. Bartes, and M. M. Ball, *Amer. J. Med. Technol.*, **37**, 99 (1971).

7. J. M. Matsen, and J. C. Sherris, *Appl. Microbiol.*, **18**, 452 (1969).

8. P. B. Smith, K. M. Tomfohrde, D. L. Rhoden, and A. Balows, *Appl. Microbiol.*, **24**, 449 (1972).

9. K. M. Tomfohrde, D. L. Rhoden, P. B. Smith, and A. Balows, *Appl. Microbiol.*, **25**, 301 (1973).

10. J. A. Washington, II, P. K. W. Yu, and W. J. Martin, *Appl. Microbiol.*, **22**, 267 (1971).

CHAPTER 26

Methods for the Rapid Identification of Enterobacteriaceae

J. LEONARDOPOULOS AND J. PAPAVASSILIOU

SUMMARY

Motility, gelatin liquefaction, and nitrate reduction of 363 strains of Enterobacteriaceae were tested in Ball and Sellers's medium (1) in comparison
to media commonly used for the same purposes. It was found that (212)
strains were motile in Ball and Sellers's medium instead of 179 in soft agar
0.5%. Ball and Sellers's medium was superior to commonly used media for
gelatin liquefaction. In addition, nitrate reduction occurred within 24 hours
in this medium.

Ullmann's (2) method for the rapid detection of indole production in
Kligler's iron agar was applied on 200 strains with very satisfactory results.
Several micromethods were used for IMViC, PPA, urease and gelatin tests.
Our micromethods for PPA, urease, indole, and methyl red (MR) tests were
very satisfactory. Numbers of strains examined were as follows: for indole,
130; for MR, 118; for VP, 90; for citrate, 100; for PPA-urease, 130; for
gelatin, 81.

INTRODUCTION

The identification of Enterobacteriaceae by commonly used methods and media requires several days or even weeks. The introduction of quick methods and especially micromethods shortened considerably the time required. During the last few years systematic attempts were made in our laboratory to select, modify and introduce new quick methods and micromethods for the rapid identification of Enterobacteriaceae. In this study we present some results obtained by two quick methods (Ball and Sellers's (1) and Ullmann's (2) methods) and by several micromethods. Some of these approaches proved to be reliable and satisfactory and are now in routine use in our laboratory.

METHODS AND RESULTS

Motility, Gelatin Liquefaction, and Nitrate Reduction

Motility, gelatin liquefaction, and nitrate reduction of 363 strains of Enterobacteriaceae were tested in Ball and Sellers's medium (1) in comparison to media commonly used for the same purposes (Table 1). In Ball and Sellers's medium, 212 (58.4%) strains were motile and 151 (41.6%), were nonmotile after 24 hours. In soft agar 0.5% (see Ref. 3) only 179 (49.3%) were motile; 166 (45.7%) strains were nonmotile, and 18 (4.9%) gave doubtful results even after 3 days of incubation. The different results obtained by the two media were more important for *Escherichia, Citrobacter,* and *Proteus* because some strains that were motile in Ball and Sellers's medium were only slightly or not at all motile in soft agar.

From 113 strains of *Proteus*-liquefying gelatin, 105 (92.9%) gave a positive result in Ball and Sellers's medium after 24 hours. The other 8 (7%) strains liquefied gelatin within 2 to 14 days. By the routine stab culture method (4), only 55 (48.6%) strains gave a positive result after 24 hours, whereas 43 strains (38%) liquefied gelatin within 2 to 3 days and 15 (13.3%) within 4 to 28 days.

Nitrate reduction test was positive after 24 hours in both Ball and Sellers's and Kauffmann's media.

Table 1 Motility and Gelatin Liquefaction in Ball and Sellers's Medium

Group	Number of Strains	Motility[a]						Gelatin Liquefaction (days)											
		Ball and Sellers's Medium			Ball and Sellers's Soft Agar 0.5% (3)			Ball and Sellers's Medium						Gelatin in Kauffmann's Medium (4)					
		M	NM	D	M	NM	D	1	2–3	4–7	8–14	15–21	22–28	1	2–3	4–7	8–14	15–21	22–28
Escherichia	100	87	13	—	80	15	5	—	—	—	—	—	—	—	—	—	—	—	—
Citrobacter	40	36	4	—	34	5	1	—	—	—	—	—	—	—	—	—	—	—	—
Klebsiella	50	—	50	—	—	50	—	—	—	—	—	—	—	—	—	—	—	—	—
Proteus	113	59	54	—	35	66	12	105	3	3	2	—	—	55	43	7	2	2	4
Salmonella	30	30	—	—	30	—	—	—	—	—	—	—	—	—	—	—	—	—	—
Shigella	30	—	30	—	—	30	—	—	—	—	—	—	—	—	—	—	—	—	—
Total	363	212	151	—	179	166	18	105	3	3	2	—	—	55	43	7	2	2	4

[a]Code: M, motile; NM, nonmotile; D, doubtful.

Indole

Ullmann's method (2) for the rapid detection of indole production in Kligler's iron agar was applied on 200 strains of Enterobacteriaceae, fermenting or not fermenting lactose but producing indole in peptone water (*E. coli I*: 120 strains, untypable coliforms 6; *Citrobacter*, 9; *Proteus*, 50; *P. coliforme*, 10; *Gloaca*, 5). A clear positive result was always obtained; no false-positive results with indole-negative strains were observed. The method proved very useful for the rapid detection of nonlactose-fermenting organisms in Kligler's iron agar. Indole-positive reaction on this commonly used medium proves that the microorganism is not *Salmonella*; usually there is no need for additional biochemical or serological tests.

Micromethods

Indole

The micromethod of Arnold and Weaver (5) gave satisfactory results with 130 strains of Enterobacteriaceae (Table 2). Sixty-three strains of *E. coli* and related organisms gave a positive reaction within 1 hour and 67 strains of *Proteus* in 2 hours. We modified (6) the method by using 0.5 ml of medium instead of 1.0 ml, and Ehrlich's reagent instead of Kovács. Results by both methods were essentially the same, but in some cases a positive test was obtained earlier by the modified method.

Single colonies of 59 nonlactose-fermenting strains taken from Mac-Conkey agar were tested for indole production both by the original micromethod and by our modification (Table 3). Results were as follows:

1. Positive tests are given with considerable delay when single colonies are used as inoculum.

2. The use of single colonies as inocula is not recommended for *Proteus* if readings must be made within 4 hours.

3. The modified method is more sensitive than the original. (See positive tests with *Proteus* within 2 hours and *P. coliforme* within 1 hour.)

Methyl Red Test

A few micromethods were used without satisfactory results (7–9). The method of Barry et al. (10) proved to be a quite reliable technique for the methyl red test in Enterobacteriaceae. Out of 118 strains examined (32 *E. coli* I, 6 *P. coliforme*, 6 *Citrobacter*, 10 *E. coli* II, 23 *Klebsiella*, 32 *Cloaca*, 3 *Aerobacter*, and 6 untypable coliforms), only 2 gave results other than the macromethod, after 24 hours.

Table 2 Production of Indole by 130 Strains of *Enterobacteriaceae*:
Micromethod of Arnold and Weaver (5)

Group or Species	Strains Examined	Time (minutes) Number of Indole-Positive Cultures			
		30	60	90	120
E. coli I	50	44	6	—	—
Untypable coliforms	8	6	2	—	—
P. coliforme	5	3	2	—	—
Proteus	67	15	8	24	20

Table 3 Indole Production by Single Colonies of 59 Nonlactose-Fermenting Strains of *Enterobacteriaceae*

Group or Species	Strains	Time (minutes) Number of Indole-Positive Cultures									
		30		60		90		120		240	
		A–W[a]	M[b]	A–W	M	A–W	M	A–W	M	A–W	M
Proteus	53	—	—	—	—	—	—	1	9	23	20
P. coliforme	6	1	1	1	4	4	1	—	—	—	—

[a]Arnold and Weaver's micromethod (5), 1 ml of medium.
[b]Modified micromethod (6), 0.5 ml of medium.

We found (11) that the same satisfactory results can be obtained by using inoculum from a nutrient agar slant, instead of single colonies, the former producing a light turbidity in the medium. This slight modification is particularly important because it permits the examination of the strains as a whole, as opposed to single colonies, which are heterogeneous and may give variable results. Furthermore, inoculum from a nutrient agar slant can be used simultaneously in several micromethods.

Voges-Proskauer Test

The method of Benjaminson et al. (12) gave positive results with 90 strains, producing acetyl methylcarbinol after 24 hours of incubation (35 *Klebsiella*, 5 *Aerobacter*, 40 *Cloaca*, 10 untypable coliforms). Attempts made to shorten the time required by inoculating heavily the medium (10^9 microorganisms/ml) did not succeed.

Citrate

The micromethods of Hargrove and Weaver (13) and of Egler and Tóth (14) were tested in comparison to Koser's (15) method (Table 4). Out of 87 strains that were citrate positive in Koser's medium, only 37 (42.5%) were positive by Egler's and Tóth's method. False-positive tests were also observed; two strains of *E. coli* I were citrate positive by this method. The micromethod of Hargrove and Weaver gave very satisfactory results. Seventy-three (83.9%) strains yielded a positive test after 90 minutes of incubation. The other 14 (16%) strains were positive after 210 minutes. No false-positive results were observed.

Table 4 Citrate Utilization by 100 Strains of *Enterobacteriaceae.*

Group or Species	Strains	Koser's Method (15) (days) 1	2	5	Hargrove and Weaver's Method (13) (hours) $1\frac{1}{2}$	$3\frac{1}{2}$	Egler and Tóth's Method (14) (hours) 6
Aerobacter aerogenes	3	2	1		3		2
Citrobacter	7	2	4	1	3	4	6
Klebsiella	22	15	7		18	4	7
Cloaca	38	21	17		33	5	20
Untypable coliforms	2	2			2		2
S. typhimurium	5	4	1		5		0
S. enteritidis	5	5			4	1	0
S. thompson	5	4	1		5		0
E. coli I	9	—	—	—	—	—	2
Par. coliforme ↑	2	—	—	—	—	—	—
Untypable coliforms	2	—	—	—	—	—	—
Total	100	55	31	1	73	14	39

Gelatin

The micromethods of McDade and Weaver (16) and of Le Minor and Piechaud (17) were tested in comparison to the routine stab culture method (4) and the method described by Ball and Sellers (1) (Table 5).

We found that Ball and Sellers' method gave the most rapid results with *Cloaca*. Out of 29 strains, 28 (96.5%) gave positive result within 48 hours, and only one strain showed gelatin hydrolysis after 3 days. McDade and Weaver's method gave the most rapid results with *Proteus*. After 1 to 3 hours, 80.7% of the strains gave a positive result. With slowly liquefying strains of *Proteus mirabilis*, this method gave negative results after 4 days, as did the routine (Kauffmann) stab culture method and the method of Le Minor et Piéchaud after 30 days of incubation at 37°C. From 52 strains of *Proteus*, 45 (86.5%) showed gelatin hydrolysis after 24 hours in Ball and Sellers's medium. The others liquefied gelatin after 2 to 12 days. Ball and Sellers's method gives very satisfactory results, mostly within 24 to 48 hours, with both rapidly and slowly liquefying gelatin strains of Enterobacteriaceae. Furthermore, motility and nitrate reduction can be tested in the same medium. For these reasons, we consider Ball and Sellers's method to be the method of choice for routine use. Quicker methods, especially McDade and Weaver's micromethod, can be used simultaneously or for the identification of genera that rapidly liquefy gelatin, as *Proteus* usually do.

PPA and Urease Tests

A micromethod has been developed based on Vassiliadis and Politi (18) medium for the simultaneous examination of PPA and urease (19). The micromethod gave satisfactory results for the PPA test; 95 strains of *Proteus* and 35 strains of *Providencia* displayed positive PPA tests within 4 hours.

The urease test was also positive for all 95 strains of *Proteus* within 4 hours. Strains of *Klebsiella* and *Cloaca* producing urease within 4 days by standard tests gave a positive result in 2 to 22 hours by the micromethod. False-positive tests were not observed (Table 6).

Table 5 Gelatin Liquefaction Test by 81 Strains of *Enterobacteriaceae*

Groups or species	No. of strains	Stab culture method (Kauffmann[4]) days													Ball and Seller's method I days								McDade & Weaver's method[16] hours		days			
		1	2	3	4	5	6	7	8	14	15	21	25	30	1	2	3	5	6	7	10	12	1–3	4–8	1	2	3	4
Cloaca	29			5	15	6	2	1							3	25	1						4	1	4	16	2(−2)	
Pr. vulgaris	11	2	4	2	1						1				10						1		10	1				
Pr. mirabilis	41	21	12						1	1	1	1	2	(−2)	35	1		2	1	1		1	32	4	1	1(−3)		
Total	81																											

Method of Le Minor et Péchaud[17]

Groups or species	No. of strains	hours		days							
		1–3	4–8	1	2	3	4	5	6	8	30
Cloaca	29	1		5	5	5	3	8	1	1	
Pr. vulgaris	11	3	4	2	1	1					
Pr. mirabilis	41	8	6	19	3	2					(−3)
Total	81										

Table 6 Urease-Production and PPA Formation by 204 Strains of *Enterobacteriaceae*: **(Micromethod of Leonardopoulos and Papavassiliou) (19)**

		Time (hours)									
		Positive Urease Tests						Positive PPA Tests			
Group or Species	Strains	$\frac{1}{2}$	1	2	4	8	9–22	2	3	4	24
Proteus	95	87	5	—	3	—	—	29	63	3	—
Providencia	35	—	—	—	—	—	—	14	16	5	—
Klebsiella	17	—	—	1	7	4	5	—	—	—	—
Aerobacter aerogenes	3	—	—	—	—	—	—	—	—	—	—
Cloaca	35	—	—	—	—	1	—	—	—	—	—
Untypable coliforms	4	—	—	—	—	—	—	—	—	—	—
E. coli 1	10	—	—	—	—	—	—	—	—	—	—
Salmonella	3	—	—	—	—	—	—	—	—	—	—
Shigella	2	—	—	—	—	—	—	—	—	—	—
Total	204										

REFERENCES

1. R. J. Ball and W. Sellers, *Appl. Microbiol.*, **14**, 670 (1966).
2. U. Ullmann, *Zentralbl. Orig.*, **208**, 3, 469 (1968).
3. R. Buttiaux, J. Moriamez, and J. Papavassiliou, *Ann. Inst. Pasteur*, **90**, 133 (1956).
4. F. Kauffmann, *Enterobacteriaceae*, 2nd ed., Munksgaard, Copenhagen, 1954.
5. W. M. Arnold and R. H. Weaver, *J. Lab. Clin. Med.*, **33**, 1334 (1948).
6. J. Leonardopoulos and J. Papavassiliou, *Acta Microbiol. Hellenica*, **16**, 99 (1970).
7. S. T. Cowan, *J. Gen. Microbiol.*, **9**, 101 (1953).
8. L. S. McClung and E. D. Weinberg, U.S. Air Force School of Aviation Medicine, Randolf AFB, Texas, Rept. 56–37, 10 pp, 1956.
9. J. Leonardopoulos, J. Papavassiliou, and O. Paniara, *Acta Microbiol. Hellenica*, **16**, 189 (1971).
10. A. L. Barry, K. L. Barnsohn, A. P. Adams, and L. D. Thrupp, *Appl. Microbiol.*, **20**, 866 (1970).
11. J. Leonardopoulos and J. Papavassiliou, *Acta Microbiol. Hellenica*, **17**, 1 (1972).
12. M. A. Benjaminson, B. C. de Guzman, and A. J. Weil, *J. Bacteriol.*, **87**, 234 (1964).
13. R. E. Hargrove and R. H. Weaver, *Amer. J. Clin. Pathol.*, **21**, 286 (1951).
14. L. Egler and M. Tóth, *Zentralbl. Bakteriol. Parasitenk. Orig. Abt.*, **189**, 294 (1963).
15. *Standard Methods for the Examination of Water and Sewage*, 9th ed., 231 (1946).
16. J. J. McDade and R. H. Weaver, *J. Bacteriol.*, **77**, 60 (1959).
17. L. Le Minor and M. Piéchaud, *Ann. Inst. Pasteur*, **105**, 792 (1963).
18. P. Vassiliadis and G. Politi, *Ann. Inst. Pasteur*, **114**, 431 (1968).
19. J. Leonardopoulos and J. Papavassiliou, *Pathol. Microbiol.*, **36**, 129 (1970).

CHAPTER 27

Easy, Economic Typing of Enterobacteria

S. W. B. NEWSOM

SUMMARY

The manufacture and use of multipoint inoculators for the Microtiter (Cooke) tray and the Repli-dish is described. These inoculators were cheap and easy to make, and together with the use of simple Perspex covers to stop airborne contamination, they enabled the screening of 1050 strains of enterobacteria. The strains were screened for their resistance to four β-lactam antibiotics and for their biochemical type; resistant strains were fully typed and examined later for β-lactamases. The enzymes that were found correlated well with the species and antibiotic resistance of the strains. The study was possible only by the use of micromethods, and the apparatus needed cost less than £ 10 stg.

INTRODUCTION

The replica-plating technique has been widely used following the introduction of an eight-prong inoculator by Garrett (1) and the velvet pad technique by the Lederbergs (2). Breach, Carr, and Codner developed a device that employed 25 rods to inoculate Durham tubes in a holder (3). Goodfellow and Grey used a similar device with small glass vials to contain the medium (4). Watt, Jeffries, and Price made an elegant automatic inoculator (5). Several commercial models are now available. The need to use small tubes and vials has been eliminated by the introduction of divided dishes and microtiter trays.

Thus, 5 years ago, when I wanted to study the epidemiology and antibiotic resistance of enterobacteria in a small cardiothoracic hospital, with minimal funds and no technical help, I thought in terms of a multiple inoculating device. It had to be simple, however, because I did not have an engineering workshop to help me. This chapter describes the production of such a device and mentions the results obtained with it.

MATERIALS AND METHODS

Equipment

The baseplate from an orbital sandpapering attachment for an electric drill (Bridges), stripped of its backing, served as the base of a 96-needle inoculator for the Microtiter trays. A Microtiter tray (Cooke) was taped to the front, and 1-mm holes were bored through the center of each well and through the baseplate. The tray was removed, and the front of the plate was layered with a 0.5-cm layer of ready-mixed fiberglass paste (Isopon). The needles (no. 11 sewing needles, with 1-mm diameter and 2-mm eyes) were inserted eye-first through the back of the plate and the fiberglass layer, then centered and leveled in the wells of another tray. The points of the needles, which projected about 1 cm behind the baseplate, were then embedded in plasticene, which held them steady while the fiberglass set. The finished device (5 years later) appears in Fig. 1. A similar device was made for inoculating the Repli-dishes (Dyos plastics).

The needles were calibrated by filling a master tray with concentrated methylene blue dye and measuring the amount carried over to the wells of

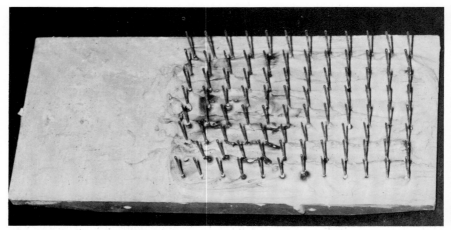

Fig. 1. Side view of 96-needle inoculator.

six other trays. The contents of each well were aspirated and diluted to 2 ml with water, and the dye was estimated in a colorimeter set at 535 nm.

Simple Perspex covers were made to hold both types of tray and to protect them from airborne contamination. Each cover contained a central slot, which exposed one row of wells at a time. Figure 2 illustrates the cover in use, while a Microtiter tray is being filled.

Bacteriological Methods

The strains of enterobacteria were isolated on horse blood or MacConkey agar. They were screened in batches for basic biochemical characters and for the ability to grow in 100 μg/ml of ampicillin, carbenicillin, cephaloridine, or cephalexin. Resistant strains were cloned and stored at $-70°C$, they were then retested against 100 and 500 μg/ml of the antibiotics and fully typed. The biochemical methods were based on Cowan and Steel (6). Micromethods did not work for urease because ammonia diffused into surrounding wells; potassium cyanide medium was dispensed in screw-cap bottles for safety. Donovan's medium (7) was used to test for motility, inositol fermentation, and hydrogen sulfide production. Repli-dishes were used for decarboxylase, phenylpyruvic acid–malonate, indole, gluconate, and Donovan's media. The fermentation and antibiotic tests were carried out in the Microtiter trays, which were sealed with the standard plate sealers.

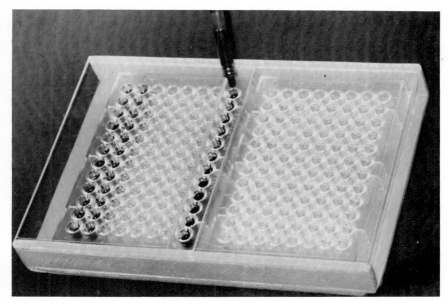

Fig. 2. Filling a Microtiter tray with medium; cover in use.

RESULTS

The needles delivered 0.0001 ml ±S.D. of 21%. The peripheral needles transferred a little more than the central ones, and the whole inoculator transferred 20% more when it was dry. The wells of the master tray had to contain equal volumes of liquid to obtain reproducible transfer.

The covers for the trays eliminated airborne contamination, during media dispensing or the filling of "master trays" with the bacterial strains. An unprotected Repli-dish, on the other hand, invariably contained one or more contaminated wells if incubated for 24 hours after filling with medium.

The results of the biochemical typing are given in Table 1. Many of the strains were classified by the primary screen, and nearly all the resistant ones were named by the final tests. The tables of Cowan and Steel adequately described most species, but the *Klebsiella* were all *K. aerogenes*, and 45% were indole positive, a feature not included in the tables. The indole-positive strains were more likely to be resistant to cephalosporins; these were "hospital" bacteria. The results of a more detailed analysis of the resistance of 350 strains appear in Table 2. These strains have been examined elsewhere for β lactamase production, and the results bear out both the typing and the resistance patterns established in this survey.

Table1. Results of the Primary Screening Test of the Ability to Grow in 100 μg/ml of β-lactan Antibiotics

Type of organism	Sensitive	Resistant	Total strains
Enterobacter and Citrobacter	25	113	138
Proteus morganii) Proteus vulgaris)	0	35	35
Proteus mirabilis	135	29	164
E. coli	307	91	398
Klebsiella	23	153	176
Others	108	31	139
Total	598	452	1050

Table 2. Resistance Patterns in 350 Strains of Enterobacteria

	Percentage of strains resistant to 100 μg/ml of the antibiotics					
	Enterobacter	Klebsiella		Escherichia	Proteus	
Antibiotics		Indole-positive	Indole-negative	E. coli	Morganii vulgaris	Mirabilis
Cb		14	11			12
A						0
Cb, A		51	66	24		38
Cb, A, C			10	24		
Cx						12
C						7
C, Cx	14			15		4
A, C, Cx	36			10	48	4
Cb, A, C, Cx	35	31	10		36	4

Cb - Carbenicillin

A - Ampicillin

C - Cephaloridine

Cx - Cephalexin

DISCUSSION

The use of trays to screen large numbers of bacterial strains for antibiotic resistance or for biochemical reactions is becoming increasingly popular. This chapter confirms that sophisticated, expensive apparatus is not required and points out that the trays allow the screening of bacteria on a scale hitherto impossible. Thus 1500 tests could be set up from a master tray in 10 minutes. They would require only 150 ml of medium and would occupy $20 \times 13 \times 8$ cm of incubator space.

Care is obviously needed to avoid cross-contamination during inoculation and to line up the inoculators. Cross-contamination was much less frequent than might be expected. Although all the strains were examined by the phenylpyruvic acid test, very few—perhaps 7 or 8—turned out to have become contaminated by *Proteus* species, which might be thought to be the most likely spreader.

The enterobacteria grew satisfactorily with the air available under the seal of the microtiter tray; more aerophilic species might not, however, and the makers now supply a "seal perforator." Alternatively, one can invert a second tray over the first one and seal both with tape.

The 350 strains that have been examined for β-lactamases contained examples of virtually every known enzyme producer, including one so far reported only from Japan. The type of β-lactamase produced, and more specifically, the susceptibility of the enzymes to certain inhibitors, has been shown to relate very well to the species of organism.[8] The study, which is still proceeding, was possible in terms of time and money only by the use of these micromethods. The results have amply justified the procedure.

REFERENCES

1. S. G. Garrett, *Trans. Brit. Mycol. Soc.*, **29**, 171 (1946).

2. J. Lederberg and E. M. Lederberg, *J. Bacteriol.*, **63**, 399 (1952).

3. F. W. Breech, J. G. Carr, and R. C. Codner, *J. Gen. Microbiol.*, **13**, 408 (1955).

4. M. Goodfellow and T. R. G. Gray, in *Methods for the Identification of Bacteria*, Part A, B. M. Gibbs and F. Skinner, Eds., Academic Press, London and New York, p. 116.

5. P. R. Watt, L. Jeffries, and S. A. Price, *ibid.*, p. 124.

6. S. T. Cowan and K. J. Steel, in *The Identification of Medical Bacteria*, Cambridge University Press, London, 1965.

7. T. J. Donovan, *J. Lab. Med. Technol.*, **23**, 194 (1966).

8. Newsom, S. W. B., Marshall, M. J. and Harris, A. M., *J. Med. Microbiol.*, (1974) in press.

CHAPTER 28

Method of Screening Cultures of Stools for Enteric Pathogens

RÜKNETTIN ÖĞÜTMAN

SUMMARY

The research was done among the population of eastern Turkey. Atatürk University Medical School, located in the eastern part of Turkey, serves as a medical reference center for a large area having approximately 5 million inhabitants. In this developing part of the country, we still can see many infectious diseases, including endemic enteric diseases. During the fall of 1972 an epidemic of enteric disease, including typhoid fever, paratyphoid A, and paratyphoid B, infections, occurred in a small area. During 2 weeks, 73 suspected cases were admitted from that area to our Infectious Diseases Department. The cases were diagnosed definitely either by positive blood culture or by positive serology.

Because of the epidemic, the local health department requested our department to join it in working in the affected area to screen the population for carriers and to locate the source of infection. The estimated population of the area to be checked was around 10,000, and no laboratory facilities were present nearby. Therefore, we organized a team and decided to use a (possibly) quicker method to screen people. However, we also used a routine laboratory method in the area to compare the reliability of the quicker method.

MATERIALS AND METHODS

The specimens were collected and cultured from the population of the area in which the cases occurred. The given number of the population of the area was 9326—4720 female and 4606 male. We could only get specimens to be cultured from the 3725 individuals, at random. Specimens were taken with specially prepared Stuart's cotton swabs. Materials used were as follows:
1. Stuart's cotton swabs.
2. Test tubes, 25×150 mm.
3. Test tubes, 15×150 mm.
4. Test tubes, 11×100 mm.
5. Pipettes.
6. Incubators for $37°C$.
7. EMB agar (dehydrated).
8. MacConkey's agar (dehydrated).
9. Selenite-F medium.
10. Slides.
11. *Salmonella* group-specific antisera.

447

PROCEDURE

In our laboratories we usually employ the routine identification method to isolate *Salmonella* species. As a rule, each specimen is cultured directly on the EMB and MacConkey's agar medium along with Selenite-F liquid medium. If there is no suspected pathogenic growth on the solid media, we transfer a loopful specimen from Selenite-F to the above-mentioned agar media. We use either biochemical or serological identification methods for suspected colonies.

In the present research we used the routine examination method, along with a screening method based on the same procedure. We used three Stuart's cotton swabs to take samples. Each person had a form and a code number, and we took 3 separate rectal samples with swabs from each person. For each swab we applied the following procedures:

1. The first rectal swab was used to streak for direct cultivation on the individual EMB and MacConkey's agar media.

2. The second swab was put in an 15×150 mm. individual test tube containing of 7 ml of liquid Selenite-F medium.

3. The third swab was put along with the 9 other specimens in a large test tube (25×150 mm) containing of 30 ml of Selenite-F medium.

All media were incubated in at 37°C overnight (approximately 18–24 hours). Following the incubation period we applied the following steps:

1. Directly cultured EMB and MacConkey's agar media were checked carefully for suspected colonies.

2. From an individual tube of Selenite-F medium, a loopful of specimen was transferred to EMB and MacConkey's agar media, using one plate for three samples. Incubation for another 18–24 hours followed.

3. From a large tube of Selenite-F medium contaning 10 specimens, a loopful of material was transferred to one-third of a plate of EMB and MacConkey's agar media. The other two parts were used for the other specimens and then were incubated for 18–24 hours.

If there were any colonies suspected of containing pathogens in the first step or in the other steps, the routine methods were used, including biochemical and agglutination methods for definite identification.

RESULTS

Table 1 and 2 give, respectively, the age and sex distribution of the population of the affected area and the age and sex distribution of the 3725 persons checked for rectal culture. The youngest person tested for enteric infection was 1 year and oldest was 83 years old. Table 2 also shows the number of identified carriers and allows us to compare the results of the individual and multiple-culturing methods.

Table 1 Age and Sex Distribution of 9326 Persons in Affected Area

Age (Years)															
0–10		11–20		21–30		31–40		41–50		51–60		61–70		70$^+$	
M	F	M	F	M	F	M	F	M	F	M	F	M	F	M	F
940	985	777	783	637	750	730	737	720	673	510	511	270	328	15	60

DISCUSSION

To evaluate a small salmonellosis epidemic in an area with a population of 9326, rectal specimen were cultured. Only 3725 persons could be checked, (1851 male and 1874 female. The average age was 42.3 years.

We cultured each specimen individually and 10 specimens together. From the cultured specimens of 3725 samples, we had 47 positive results with individual culturing and 42 positive results with the multiple-sample culturing method. This indicates that the single or individual-specimen culturing method has very little superiority over the other approach. Using the multiple-sample method, we believe that we can cut expenses of media and time a great deal, especially for large-scale screening projects.

Table 2 Age and Sex Distribution of 3725 Persons Checked, and Positive Findings for Carrier

| | Age Groups and Sex | | | | | | | | | | | | | | | | | |
| | 1–10 | | 11–20 | | 21–30 | | 31–40 | | 41–50 | | 51–60 | | 61–70 | | 70$^+$ | | Total | |
	M	F	M	F	M	F	M	F	M	F	M	F	M	F	M	F	M	F
Number of cultures	385	400	351	355	304	350	288	298	270	282	191	153	37	63	5	13	1811	1914
Number of positives (individual)	4	5	—	3	8	7	5	2	6	—	4	1	—	2	—	—	27	20
Number of positives (multiple)	3	6	1	2	6	6	5	—	4	1	4	1	1	1	—	1	24	18

Primary Amoebic Meningoencephalitis: Agar Plate Method for Rapid Detection and Identification of Naegleria–Hartmannella Groups of Amoebae

S. R. DAS

SUMMARY

The use of nonnutrient agar with a suitable edible bacterium, such as *Aerobacter aerogenes* or *Escherichia* has led to a satisfactory method of culturing *Naegleria–Hartmannella* groups of pathogenic free-living amoebae from infected tissues, soil, sewage sludge, fresh water, and other substrates. A culture method for growing small, free-living amoebae on a thin film of nonnutrient agar has facilitated the study of various stages of nuclear division in these amoebae. In the genus *Naegleria*, the amoebae have polar masses and interzonal bodies during mitosis. The temporaty flagellate stage has two flagella. In the genus *Hartmannella*, the resting nucleus contains a single Feuelgen-negative nucleolus. During mitosis, the nucleolus disappears and a spindle with chromosomes arranged as an equatorial plate is formed. No temporary flagella are produced.

INTRODUCTION

The discovery that the *Naegleria–Hartmannella* groups of small, free-living amoebae can cause fatal human and animal meningoencephalitis has changed the whole concept of amoebiasis. *Entamoeba histolytica* is no longer regarded as the only amoeba pathogenic to man.

Extensive studies carried out by B. N. Singh at Rothamsted Experimental Station, England, from 1941 to 1952 led to the use of nonnutrient agar or silica jelly and a suitable edible bacterium, such as *Aerobacter* sp., for the isolation and culture of small, free-living amoebae and amoeboid organisms from soil and other substrates and for the enumeration of their numbers in these substrates. A rational system of classification of small, free-living amoebae, based on their nuclear structure, nuclear division, and other characters and possible phylogenetic relationship, was also developed by Singh (10,11). It was inconceivable in the early 1950s, even to imagine that free-living amoebae might cause disease in man.

In a series of papers since 1958, C. G. Culbertson and his colleagues (7) have demonstrated conclusively that trophozoites of *Hartmannella* (strain A-1), contaminant of monkey kidney cell cultures, and other strains of the same genus isolated from soil and other sources, when given intranasally to mice and monkeys, invaded the brain directly through olfactory route and caused fatal meningoencephalitis. These workers also noted that a similar event may be produced in man by the entry of amoebae through the nasal route during swimming in fresh water contaminated with pathogenic *Hartmannella*. Fowler and Carter (9) first reported four fatal human cases of acute pyogenic meningitis caused by free-living amoebae in Australia. They thought that these cases were probably due to *Hartmannella* (*Acanthamoeba*). Since that time, more than 60 human primary amoebic meningoencephalitis cases have been reported from different parts of the world.

Hartmannella was believed to be the causative agent of human primary amoebic meningoencephalitis cases before 1968. Carter (4) in Australia and Butt, Baro, and Knorr (3) and Culbertson, Ensminger, and Overton (8) in the United States reported the isolation of an amoebo-flagellate from human cases of meningoencephalitis. They called this amoeba *Naegleria* sp. Since these publications, amoebae isolated from human cases of meningoencephalitis have turned out to be *Naegleria* sp. It is now generally believed that *Hartmannella* infections reported before in man may have been caused by *Naegleria*.

RECENT WORK

The epidemiology of human meningoencephalitis caused by *Naegleria* sp. is yet not properly understood. Recently Singh and Das (13) found that *Naegleria* sp., which is pathogenic to mice, is commonly present in sewage sludge samples of Lucknow. They demonstrated that intranasal inoculation of mice with specimens of the temporary flagellate stage of pathogenic *Naegleria* also causes meningoencephalitis and death of mice (14). They have suggested (14) that the flagellate stage of *Naegleria* is responsible for causing meningoencephalitis in man. Pathogenic *Hartmannella* does not produce a flagellate stage; therefore, it has little chance of entering the nose during swimming in fresh water.

Anderson and Jamieson (1,2) in Australia have reported the isolation of pathogenic *Naegleria* from a sample of tap water taken from the house of a fatal case of human meningoencephalitis. Ten strains of pathogenic *Naegleria* have also been isolated by these workers from a domestic water supply. In addition, a sample of surface soil revealed the presence of this amoeba. It may be pointed out that there is more chance of the prevalence of primary amoebic meningoencephalitis in countries having hot climates than in countries that are very cold for most of the year, because pathogenic *Naegleria* grows better at 37°C than at lower temperatures. During the rainy season, when the temperature is high in tropical countries, children often play in muddy water and water contaminated with sewage sludge. There is a likelihood that pathogenic *Naegleria* flagellate will enter some of these children through the nose. A proper investigation of the etiology of human meningoencephalitis cases in hot countries will probably reveal that amoebic meningoencephalitis is less rare than had been assumed.

The use of nonnutrient agar and a suitable edible bacteria, such as *Aerobacter aerogenes* or *Escherichia coli*, developed by Singh, is the method of choice for the isolation and culture of *Naegleria–Hartmannella* groups of amoebae from human and animal cases of meningoencephalitis, soil, fresh water, sewage sludge, and other substrates. Small free-living amoebae do not require any nutrient other than that provided by bacteria. This method is being used by different workers throughout the world who are engaged in the study of pathogenic, small, free-living amoebae causing meningo-encephalitis.

Naegleria

The genus Naegleria Alexeieff emend. has been included in the family Schizopyrenidae (10) and is defined as follows. Polar masses are formed, and Feulgen-negative interzonal bodies are present during late stages of nuclear division. Temporary flagella are produced. The flagellate stage has two flagella, and no division takes place in this stage. In the genus *Didascalus*, polar masses without interzonal bodies are present during nuclear division (10). Temporary flagella are produced. The flagellate stage has two flagella and no division takes place in this stage. Pathogenic *Naegleria* (Culbertson strain HB-1) has been named by Singh and Das (12) as *N. aerobia* sp. nov. on the basis of aerobicity of the organisms. It differs from nonpathogenic *N. gruberi* with respect to the following characteristics:

1. Under similar cultural conditions and with similar food supply, the trophic, cystic, and flagellate forms of *N. aerobia* are smaller than those of *N. gruberi*.

2. *N. aerobia* grows best at 37°C and cannot tolerate 0.5% NaCl incorporated in nonnutrient agar. *N. gruberi* grows best at 24 to 25°C and can tolerate 0.5% NaCl.

3. *N. aerobia* cysts have no pores, and there is an thick outer gelatinous layer; during excystment, some digestion of the cyst wall occurs. *N. gruberi* cysts have one or more pores, and during excystment the amoeba escapes through one of them.

4. *N. aerobia* and *N. gruberi* are serologically distinct, as judged by immobilization reaction.

Carter (5) has named the amoebo-flagellate from human cases of meningoencephalitis *N. fowleri* sp. nov. This amoeba is very similar to *N. aerobia* in cultural requirements, in trophic and cystic morphology, and in the mode of excystment. Since Carter did not observe interzonal bodies, his amoeba should be called *D. fowleri*. If it is found that Carter's amoeba possesses interzonal bodies during nuclear division, *D. fowleri* should become the junior, nonvalid synonym of *N. aerobia*.

Hartmannella

The genus *Hartmannella* Alexeieff emend. (10) has been included in the family Hartmannellidae Volkonsky, 1931 emend. (10) and is defined as follows. The resting nucleus contains a single Feulgen-negative nucleolus. During mitosis the nucleolus disappears, and a spindle with chromosomes arranged as an equatorial plate is formed. No temporary flagella are produced. Singh and Das (12) found that pathogenic *Hartmannella* belongs to two distinct species based on cystic character. The cysts of *H. rhysodes* (10) consist of two wrinkled walls, irregular in outline. The outer wall has folds and ripples and is often loosely applied to the inner wall, which is irregularly stellate, with truncated rays, or irregularly polyhedral. The two walls are in contact at points along the inner wall, where pores, plugged with structureless substance, are present. *H. culbertsoni* (12) cysts are rounded or oval and have two walls. In some cysts the outer wall is nearly circular, and in others it is irregular in outline. The cysts are perforated with one or more pores or opercula plugged by a structureless substances. Some serological relationship has been found to exist between *H. rhysodes* and *H. culbertsoni.*

The system of classification proposed by Singh (10) and Singh and Das (12) has been accepted by Culbertson (7) Chang (6), Butt, Baro, and Knorr (3), and many others studying pathogenic and nonpathogenic, small, free-living amoebae.

REFERENCES

1. K. Anderson and A. Jamieson, Primary Amoebic Meningoencephalitis, *Lancet*, **1**, 902 (1972).

2. K. Anderson and A. Jamieson, Primary Amoebic Meningoencephalitis, *Lancet*, **2**, 379 (1972).

3. C. G. Butt, C. Baro, and R. W. Knorr, *Naegleria* (sp.) Identified in Amoebic Encephalitis, *Amer. J. Clin. Pathol.*, **50**, 568 (1968).

4. R. F. Carter, Primary Amoebic Meningoencephalitis: Clinical, Pathological, and Epidemiological Features of Six Fatal Cases, *J. Pathol. Bacteriol.*, **96**, 1 (1968).

5. R. F. Carter, Description of a *Naegleria* sp. Isolated from Two Cases of Primary Amoebic Meningoencephalitis, and of the Experimental Pathological Changes Induced by it, *J. Pathol.*, **100**, 217 (1970).

6. S. L. Chang, Small Free-Living Amoebas: Cultivation Quantification, Identification, Classification, Pathogenesis, and Resistance, in *Current Topics in Comparative Pathology*, vol. 1, Academic Press, New York and London, 1971.

7. C. G. Culbertson, The Pathogenicity of Soil Amebas, *Ann. Rev. Microbiol.*, **25**, 231 (1971).

8. C. G. Culbertson, P. W. Ensminer, and W. M. Overton, Pathogenic *Naegleria* sp.—Study of a Strain Isolated from Human Cerebrospinal Fluid, *J. Protozool.*, **15**, 353 (1968).

9. M. Fowler and R. F. Carter, Acute Pyogenic Meningitis Probably due to *Acanthamoeba* sp.: A Preliminary Report, *Brit. Med. J.*, **2**, 740 (1965).

10. B. N. Singh, Nuclear Division in Nine Species of Small Free-Living Amoebae and its Bearing on the Classification of the Order Amoebida, *Phil. Trans. Roy. Soc. London. Ser. B*, **236**, 405 (1952).

11. B. N. Singh, *Culturing Soil Protozoa and Estimating Their Numbers in Soil*, Butterworths, London 1955.

12. B. N. Singh and S. R. Das, Studies on Pathogenic and Nonpathogenic Small Free-Living Amoebae and the Bearing of Nuclear Division on the Classification of the Order Amoebida, *Phil. Trans. Roy. Soc. London, ser. B*, **259**, 435 (1970).

13. B. N. Singh and S. R. Das, Occurrence of Pathogenic *Naegleria aerobia, Hartmannella culbertsoni*, and *H. rhysodes* in Sewage Sludge Samples of Lucknow, *Current Sci.*, **41**, 277 (1972).

14. B. N. Singh and S. R. Das, Intranasal Infection of Mice with Flagellate Stage of *Naegleria aerobia* and its Bearing on the Epidemiology of Human Meningoencephalitis, *Current Sci.*, **41**, 625 (1972).

LIST OF ABSTRACTS

The following contributed papers are available only in abstract form.

G. van der Ploeg, Ph. van Elteren, and J. Creemers-Molenaar, Hemoglobin reduction method in testing sensitivity of micro-organisms (abstract A 33).

Miles G. Hossom, Micro tube dilution susceptibility test (abstract A 34).

B. J. Harrington, D. J. Hansson, J. F. Dooley, and J. E. McKie, Autobac 1–A three-hour, automated antimicrobial susceptibility system. III. Clinical evaluation (abstract A 35).

Stanely Scher, Developing alternative strategies for automating microbiological techniques (abstract A 7).

Z. G. Bánhidi, Twelve channel continuous density recorder and its applications (abstract A 41).

S. K. Biswas, Use of gradocol membrane filter in rapid diagnostic bacteriology (abstract A 41).

A. Lundin, C. Blondell, and A. Thore, Rapid extraction and assay of adenine nucleotides from microorganisms (abstract B 63).

A. Thore, S. Ánséhn, S. Bergman, and A. Lundin, Detection of bacteriuria by luciferase assay for ATP (abstract B 64).

E. Chapelle, G. Picciolo, and V. Nusbaum Bush, ATP measurement by firefly luciferase for the rapid detection of bacteria (abstract B 65).

L. Ewetz, Determination of microbial material in water and urine by the luminol reaction (abstract B 66).

R. Freake, L. Strenkoski, J. Reback, C. Wiemeri, A novel, convenient, combined rapid biochemical and confirmation quantitative test system for the detection of urinary tract infections (abstract B 67).

Jerry J. Tulis, Radiometric approach to microbial detection and antibiotic sensitivity testing (abstract B 68).

Daniel Amsterdam and Marian W. Richter, Discrimination of generic variants of E. coli by differential light scattering (abstract B 69).

Jan. O. Kjellander, A computer system for clinical microbiology (abstract C 92).

Thomas J. Sgouris, Rapid detection of diphtheria and tetanus antibodies by counterelectrophoresis (abstract A 46).

W. I. Hopkinson, D. F. Gibbs, and E. J. Bennet, Agglutination detection by automated methods (ADAM) (abstract A 47).

E. Engelbrecht, An improved fixation technique for antigen carrier- and M-erythrocytes of reduced own antigenicity (abstract A 48).

Dan Danielsson and Göran Kronvall, A rapid slide agglutination method for the serologic identification of neisseria gonorrhoeae with gonococcal antibodies absorbed to protein A-containing staphylococci.

R. E. Trotman, The application of automatic methods to hospital diagnostic bacteriology (abstract B 70).

Tibor Illéni, Georg Görtz, C.-G. Hedén, and Lars Wegstedt, The optical measurement of inhibition and growth zones in bacterial films on agar media (abstract B 72).

Jerzy Borowski, Anna Boron-Kaczmarska, and Danuta Dzierzanowska, Application of the multiinoculator in laboratory practise (abstract B 74).

Jeptha E. Campbell, Spiral plating and counting of bacteria (abstract B 75).

M. T. Bomar, and S. Schmid, Control of the bacterial breakdown of cellulose—as a limiting carbon source substrate—by measurement of carbon dioxid production under aerobic conditions (abstract B 76).

Juan-R. Mor, Automatic assay of microbial growth (abstract B 77).

I. A. Martin and N. D. Harris, The rapid determination of optimum growth temperature (abstract B 78).

Mohamed A. Fouda, Automatic exponential dosing device (abstract B 79).

Walther Heeschen, The automated estimation of bacteriological keeping quality of milk by the enzymatic determination of pyruvate (abstract A 52).

M. A. Pisano, W. S. Oleniacz, N. H. Rosenfled, and R.L. Elgart, Detection of microorganisms by chemiluminescence (abstract A 53).

K. Bryn, J. C. Ulstrup, and F. C. Storner, Rapid Voges-Proskauer test with acetate (abstract A 56).

T. K. Niemelä and H. G. Gyllenberg, A numerical classification technique to provide reference systems for computer-assisted identification (abstract B 85).

H. G. Gyllenberg, J. S. Niemi, Hannele Jousimies, Annele Hatakka, and T. K. Niemalä, Two examples of development on specific computer-assisted identification procedures (abstract B 86).

Stephen P. Lapage, Probabilistic identification of bacteria (abstract B 87).

J. D. Piguet, Mini-computer assisted identification of gram-negative bacteria (abstract B 88).

INDEX

Access to a computer, 270
  breakdown, 270
  "local" mode, 270
  modem (modulating and demodulating),
    270
  transmitting, 270
Acetoacetate, 341
Acetyl methylcarbinol, 428
Achromobacteriaceae, 336
Aerobacter, 427, 428
Agar ribbon, *see* Agar strip
Agar strip, 18
  cutter, 18
  of nutrient agar, 24
  preparation and transport of, 24
  technique, 22
Alizarine, 322
Amoebic Meningo encephalitis, 451
  rimary, 451
Amojell-mineral oil mixture, 355, 358
Analysis, sequential, 51
Animal cells, 12
Antibiotics, impedance change and, 88
Antibiotics resistence testing, 28, 70
  bioautography of, 29
  determination of, in body fluids, 30
Antisera, 35
  inhibition of growth by, 95
API system, 393, 395, 397, 400, 407, 409,
    411, 412, 415, 419
  gram negative bacteria, 395
  numerical taxonomy, 401
Arboviruse s group B, 309
  antibiotics to, 309
Arizona, 323, 409, 411, 412
Aspergilli, 336
Autoline, 18
Automatic Identification, 205. *See also*
    Identification
Auxotab, 393, 395, 403
  gram-negative bacteria, 395
Auxotyping, 31

BAC-DATA, 279
  antibiotic sensitivity report, 280
  Antibiotic Susceptibility Summation, 286
  bacteriologic trend report, 282
  Bulletin, 287
  monthly trend reports, 282
  number and percentage of organism, 282
  prevalent patogens report, 287
  Research Report, 288
  Survey of methodology, 290
Bacterial film, 33
Bacterial protein patterns, 195
Bacterial pyrograms, 172, 184
  characteristic patterns, 184
  computer matching, 187, 189
  degree of correspondence, 188
  fingerprint regions, 173
  influence of anaerobic conditions, 172
  influence of culture medium, 172
  reproducibility, 172, 185
  visual comparison, 173
Bacteriogram, 34
Bacteriological Data System, 293
  BACTLAB system, 295
  cost, 299
  laboratory record, 299
  postoperative infections, 293
  statistics, 299
BACTLAB-system, 295
Ball and Sellers' medium, 423
Bioautography of antibiotics, 29
Biochemical functions, 51
  network of, 51
Biochemical identification, 391
Blood, 129
Blood culture, 447
  positive for enteric pathogens, 447
Bromthymol blue, 31

Calorimeter, 106
Candida mycoderma, 341
Capillari tubes, 349, 355

carbohydrate fermentation test, 351
catalse activity, 349
  litmus milk, 351, 355, 362
Cell parameters, 242
  time after infection, 242
  trend parameter, 242
Channel plate, 49
Citrate, 429
Citrobacter, 323, 409, 411, 412, 418, 425,
      427
Classification, classify, 219, 227, 241
  cell alteration, 227
  cell changes, 227
  by identification, 219
  molecular structure changes, 227
  virus-infected cells, 227
Cloaca, 427, 428, 430
Colonies, 321, 326, 329
  color masters, 321, 326
  morphotyper, 326
  patterns, 329
  transfer experiments, 329
Colony detection, automatic, 46
Colony identification, automation of, 3
Complement, 94
  titration of, 94
Computer, 103, 257, 259
  access to, 270
  analysis, 147
  application of, to medical microbiology,
      257
  "byte" machines, 259
  compatibility, 51, 52
  integral, 103
  misconceptions about, 257
  "word" machines, 259
Computer-aided analysis, 229, 230
  class properties, 230
  classification, 230
  scanning microscope, 230
  digitized gray values, 23
Computer clustering, 195, 196
  of electrophoretic protein patterns, 195
Conductivity, 77
Conductometric measurement, 77
Congo red, 321, 323, 326
Culturing Methods, 449
  individual, 449
  multiple, 449
Curie point pyrolysis, 168, 170
  automatic sample exchanger, 170
  Curie temperature, 171
  HF power supply, 170
  pyrolysis reactor, 171
  sample preparation, 170
  technique, 168, 170, 184

temperature rise time, 171
Cytochrome oxidase, 376

Data analysis, 134
Data system, bacteriological, 293. *See also*
      Bacteriological data system
Day book, 265
Diagnostic keys, 59
Diagnostic kits, 393, 395, 397
  API, 393
  Auxotab, 393
  Enterobacteriaceae, 393, 397
  Enterotube, 393
  gram negative bacteria, 395
Diagnostic Microbiology, 153
Dielectric constant, 77, 86
Diffusion centers, 18
  handling of, 28
Dilution plating, 41
Diplococus, 125, 288
  pneumoniae, 135
  resistance to tetracycline, 288
Distribution of the population, 448
  age, 448
  sex, 448
Dyes, 321, 322, 323
  concentration in agar, 321
  inhibiting effect of, 323
  staining effect of, 323

Escherichia, 125, 409, 411, 425
  coli, 142, 323, 325, 328, 337, 427, 429
  detection of, 83
  indole production, 325
Electronic zone analyser, 289
Electrophoretic protein patterns, 195
  computer clustering of, 195, 196
Enteric diseases, 447
  carriers for, 447
  endemic, 447
  epidemic, 447
  to screen people for, 447
Enterobacter, 288, 409, 411, 412, 419
  aerogenes, 412
  Herbicola-lathyri, 288
  sensitivities to ampicillin, 289
Enterobacteria, 437
  antibiotic resistens of, 439
  biochemical type of, 437
  β-lactamases, 437
Enterobacteriaceae, 34, 371, 395, 397, 407,
      409, 411, 412, 418, 419, 421, 423
  diagnosis, 397
  identification of, 397
  rapid identification of, 371, 374, 421
  standard biochemical tests, 404

Enterococci, 31
Enterotube, 393, 395, 397, 398, 401, 418
  gram negative bacteria, 395
  improved, 397, 409, 411, 414, 415, 418,
    419
  numerical taxonomy, 395
Enterotube Roche®, 389
Enzymatic analysis, 333
Enzymograms, 35

Feature extraction, 233
  binari image, 236
  cytoplasm granularity, 233
  digital filter, 233
  histograms, 236
  positive differences, 233
  textural gray values, 233
  two-dimensional filtering, 233
Fermentation, 340
  enzymatic analysis in, 340
Field Desorption Mass Spectrometry, 155
Field Ionisation, 155
File Interrogation, 267
  data management, 267
  conditions, 268
  control functions, 268
  monitored, 269
  pattern recognition, 269
  routine reports, 269
  surveillance, 269
Filter paper, bands of, 28
Fluorescent antibody technique, 33
Fluorometer, 54, 55, 57
  absorbance measurement, 57
  automatic spectrophoto-, 54
  excitation, 57
  fluorescence emission, 57
  quartz fiber optics, 55
Flying spot scanner, 5
FORTRAN, 250

Gaschromatography, 130
  profiles, 142
Gas-detection, 344
  production, 355, 358
Gas phase control, 20, 22
Gelatin, 28
  liquefaction, 423, 425, 430
Gel techniques, 17
  diffusion in, 17
  fresh surface of, 23
  handling and analysis, 17
  strips, 19
  thickness, 23
Genetics, of bacteria, 7
Germ counts, 42

Glass strips, 19
Glycerol, 341
Glycerokinase (GK), 341
Gradients, 29
  of (long) concentration, 29
Growth curve, bacterial, 82
Growth of microorganisms, 77

Haemophilus. 125
  influenzae, 137
Heat profiles, 103
Henry Ford Hospital, 239
Hydrogen sulfide detection, 377
Hydroxybutyrate (3), 341
  dehydrogenase (3-HBDH), 341

Identification, 101, 123, 335, 389, 391, 448
  adjustment of grouping, 220
  automatic, 205
  of bacteria, 101, 193
    by numerical analysis of electrophoretic
      protein profiles, 193, 195
  biochemical, 391, 448
  clinical material, 221
  by correlation coefficients, 217
  of Enterobacteriaceae, 389
  Euclidean distances, based on, 213
  by gaschromatography, 123
  grouping and feedback, 219
  matrix, 206
  by metabolic patterns, 88
  of microorganisms, 71, 123, 335
    in clinical specimen, 127
    computer assisted, 201
    by reagent impregnated teststrips, 371
  by probabilities, 210
  rapid, 71, 421
  serological, 448
  states, 208
  of Streptomyces, 205
Identification systems, 375
  stability of, 375
Impedance measurement, 61, 77, 84
  bridge, 77, 79
  capacitance, 84
  cells, 79
  conductivity, 84
  electrodes, 79
  frequencies, 84
Improved Enterotube, 407, 409, 411
IMViC, 423
Indole, 423, 427
  production detection, 376
Inhibition zone, 29
  measurement of, 29
Inhibitors, 339

Inoculation, 26
Instrumentation, 105
International *Streptomyces* Project (ISP),
    205
Isoelectric focusing, 35

Kauffmann's media, 425
Kirby-Bauer technique, 289
    modification of, 289
Klebsiella, 169, 177, 191, 322, 323, 409,
    411, 412, 427, 428, 430, 441
    pneumoniae, 31, 412
    aerogenes, 441
Kliger's iron agar, 423

Linear scanning, 18
Long concentration gradients, 29
Lowest unique identifier, 259
Luciferin/luciferase, 30
Luminol, 30
Lymphocytes, 98
    impedance change and, 98
Lysine, 30

Mac Conkey agar, 427
Main key, 259
Malate-dehydrogenase, 343
Malonate utilization detection, 377
Mandatory reports, 264
Manzini technique, 33
Mark-sense, 279
    input data form, 279
Mass Spectrometry, 155
Media, bacteriologic, 80
Medical data bank, 299
Medium, 447
    EMB agar, 447
    Mac Conkey's agar, 447
    Selenite-F, 447
Metabolic functions, 51
    indicators, 30
    pathways, 51
Metabolism, of microorganisms, 77, 336
Metabolite detection, 135
Methyl red test (MR), 423, 427
Methionine, 30
Microbiological Analysis, 30
    films, 33
    genetics, 31
Microbiology, clinical, 105
Microcalorimetry, 103
Micromethods, 423, 427, 437
Microtiter plates, 353, 354
    carbohydrate fermentation, 358, 369
    culture transfer, 357
    H_2S production, 363

IMViC tests, 349, 351, 355, 358
    indole test, 355
    inoculate solid, semisolid, liquid media,
        357
    mass transfer, 349, 351
    methyl red, 355
    miniaturized triple sugar iron (TSI) agar,
        355, 361
    mini agar shants, 349
    mini fermentation tubes, 349
    motility, 349
    pinhead device, 356
    pinpoint device, 356, 363
    Sellers' medium, 355
    semisolid media, 349
    sulfide indole motility (SIM), 355, 363
    Voges-Proskauer, 355
Microtiter system, 349. *See also* Microtiter
    plates
Microtiter tray, 439
Miniaturized culture vessels, 351
    divided petri-dishes, 351
    ice cube trays, 351
    Microtiter plates, 351
    trays, 351
        of glass, 351
        of metal, 351
        of plastic, 351
Mobile caravan laboratory, 309, 312
    advantages, 311
    equipment, 311
Molecular biology, 7
Motility, 423, 425
Multiple-inoculation devices, 349, 351, 352,
    355, 356, 357, 363
    bolts, 351
    capillary tubes, 351
    needles, 351
    Pasteur pipettes, 351
    piggy-back, 351
    pinhead multipoint indicator, 355, 356,
        363
    pins, 351
    plastic stamps, 351
    sterilization, 357
    syringes, 351
    velveteen, 351
Multipoint inoculator, 437
Mutants, 31
    genetic mapping of, 7
Mutant selection, automation of, 3
Mycobacteria, 125, 169, 177, 191
    from infected patients, 142

Naegleria-Hartmannella, 451
    identification of, 451

rapid detection of, 451
NASA, 17
Neisseria gonorrhoeae, 98
  detection of, 98
Neisseria meningitidis, 36
Niacin production detection, 377
Nitrate reductase detection, 376
Nitrate reduction, 423, 425
Nitrophenylglycerol (beta), 46
Nucleric acid, measurement, 238

ONPG, 378
Optical mark page reader, 295
Optical mark reading, (OMR), 266
Optical mark sense sheets, 295
Optical scanning system, 17, 22
Optional reports, 264
Optoelectronic colony identification equipment, 45
Organisms, detection of slow-growing, 70
Ouchterlony tests, 34

Parallel Multiple Analysis, 47, 52. *See also* PAVIAN
Pasteurella pestis, 321
Pathogens, 5
  growth, 5
  morphology of colonies, 5
Patho Tec, 393, 395, 397, 400, 407, 409, 411, 412, 418, 419
  gram negative bacteria, 395
  improved, 414
Pattern recognition, 227
Penicillia, 336
Penicillium chrisogenum, 337
Pipetting devices, 52
  automatic, 54
  motor operated, 53
  overflow pipettes, 52, 54
Phenolketonuria, 30
Phenol red, 31
Phenylalanin deaminase detection, 376
Plate counting, 81
  glass beats and, 81
Poliomyelitis virus, 229
  computer-aided analysis, 229
  morphological relationships, 229
  structural or textural differences, 229
  virus multiplication, 229
    kinetics of, 229
Polyacrylamide gel electrophoresis, 195
Porous ceramic roller, 26
PPA, 423, 430
Preservatives, 340
Procaryote cells, 51
Proteus, 125, 323, 409, 412, 425

coliforme, 427
mirabilis, 412, 430
morgani, 412
rettgeri, 412
Providencia, 430
Pseudomonas, 34, 125, 322
  aeruginose, 322, 323
  pseudomallei, 321
Pyrolysis, 167, 168, 169
  capillary columns, 168
  column degradation, 167, 169
  full automation of, 168
  reproducibility, 167, 169
Pyrolysis gas-liquid chromatography, 167
Pyrolysis mass spectrometry, 168, 182
  analysis rate, 169, 184, 191
  combination mass spectrometer signal average, 183
  quadrupole mass analyser, 182
  low-voltage election impact ionization, 182

Rapid Automated Identification, 153
Rapid Identification, 421
  of microorganisms, 71
R/B tubes, 393, 395, 397, 402
  gram negative bacteria, 395
"Ready-to-use" test system, 389
Record file, 259
  structure of, 259
    data areas, 259
    data fields, 259
Redundant data, 59
Repli-dishes, 439
Reproducibility, 375
  in bacteriological system, 375
Rhodopseudomonas sphaeroides, 341
Right of access, 271
  access level, 271
  authorized users, 271
  password, 271

Salmonella, 125, 173, 322, 328, 409, 411, 412, 427, 447
  antisera, 447
    group specific, 447
  computer matching of, 173
  species, 448
  typhimurium, 135
Scanning microscope, 238
Screening cultures, 447
  for enteric pathogens, 447
  of stools, 447
Screening people, 447
  quicker method, 447
  at random, 447

routine laboratory method, 447
Screening projects, 447
  large scale, 449
  at random, 447
Security, 230
  additional precautions, 230
  copy of the file, 230
Sensitivity test, 42
Sequential analysis, 51
Serology, 33, 447
  positive for enteric pathogens, 447
Serological surveys, 313
Serratina, 125, 286
  marcescens, 137
Serum analysis, 142
  by gas chromatographic profiles, 142
Shigella, 322, 323, 411
  boydii, 411
  flexneri, 411
  dysenteriae, 411
  sonneii, 411
Staphylococcus, 125, 325, 326
  within the genus of, 325
Staphylococcus aureus, 137, 322, 323, 325,
    327, 354
  characterization of, 368
  chronic infections, 326
  effects of (42) dyes, 368
  enterotoxins A, B, C, 356, 365
  food poisoning outbreak, 352
  isolates, 352, 356, 365
  mannitol salt agar, 355
  microtiter-hemagglutination inhibition,
    356
  serotypes, 326, 329
  staphyloccous medium no. 110 with egg
    yolk, 355
  tellurite-polymyxin-egg yolk agar, 355
  turkey meat, 352, 365
  turkey rolls, 352, 365
Stool examinations, 325, 328
Streptococcus, 125, 169
  faecalis, 137, 185, 189
  mutans, 177, 185
  piogenes, 94
    detection of, 96
  sanguis, 176, 189
Streptococcus group D, 322
Streptolysin O, detection of, 94
Stuart's cotton swab, 447
Subcellular particles, fractionation of, 339
Substrate tape, 41
  culivation tape, 41
  flexible carrier tape, 41
Sugars, 326, 327
  role of, 327
  glucose, 326

Supervised learning, 231, 232, 236, 237, 238
  classification rule, 232
  classified, 238
  correlation, 232
  feature vectors, 231
  likelihood, 232
    average, 232, 236
    individual, 237
    ratios, 237
  linear combination, 232
  nonparametric procedure, 232
  optimization, 237
  random variables, 237
  statistical distribution, 232
    properties, 237
  unclassified, 236
  weights, 232, 237
Surface sampling, 34

Taxonomie, 58
  multi variate statistics, 59
  numerical, 59
  species, 59
Tetrazolium chloride (TTC), 322, 323, 328
Threonine, 30
Tissues, 129
Toxinograms, 35
Tryptophane, 30
Turkey, 447
  Atatürk University Medical School, 447
  Infectious Diseases Department, 447

Ullmann's method, 423
Urease test, 430
Urinari tract infections, 28

Vaccination, 309
  with the 17D strain, 309
Viable organisms, estimation of, 69
Vibrio, 169
Viremia, 309
Virus, 125, 128, 142
  herpes, 125, 128, 142
  influenza, 125, 128, 142
  vesicular stomatitis virus (VSV), 125, 128,
    142
Virus-host relationship, 240
  kinetics, 240
  recognition and classification, 240
    of the infected cell, 240
  statistical information, 240
Voger-Proskauer reaction detection, 376
Voges-Proskauer Test (VP), 423, 428

Water blue, 323, 328

Yellow fever epidemic, 309